THE MANAGEMENT ANTHOLOGY SERIES

Edited by
Marlene G. Mayers, R.N., M.S., F.A.A.N.

Theme I.

The Organization: People and Structures

Organizational Theories and Structures
Organizational Units and Groups
Individuals in Organizations
Intraorganizational Conflict
Organizational Change

Theme II.

Management Functions

Communicating
Planning
Organizing
Staffing
Controlling
Leadership in Nursing

Theme III.

Product and Service Cost Effectiveness

Philosophy and Goals
Quality Measurement Methods
Maintaining Cost Effectiveness
Management Goals

Theme IV.

Employee Growth and Satisfaction

Motivation
Job Enrichment
Development Through Education
Development of Nurse Managers
Personal Growth

Theme V.

Organizational Security and Longevity

Organization-Environment Relationships
Social Policy and Nursing Services
The Women's Movement and Nursing
New Human Services and Technologies
Labor Relations in Nursing
Legal Protection of Patients and Nurses
The Politics of Nursing

Several volumes were in preparation when Organization-Environment Relationships was published.

Organization-Environment Relationships

Organization-Environment Relationships

Edited by

Janet M. Kraegel, M.S.N., Ph.D.
Health Care Consultant
Patient Care Systems, Inc.
Milwaukee, Wisconsin

The Management Anthology Series
Theme Five: Organizational Security and Longevity
Nursing Resources, Inc.

Introduction to the Management Anthology Series

As a nurse administrator or manager, have you often wished you could turn to your bookcase and select just the right book for the problem at hand? Or that you could talk with another nurse administrator who has faced a similar situation? Your time is limited, your problem is volatile, and the pressure merciless. Yet your bookcase contains no substantive reference source on current theories, thinking, or management methods.

Or if you are aspiring to become a nurse administrator or manager, do you wonder what you should read as part of your career development program? You may have scanned some management textbooks, only to find that, of necessity, they touch upon each subject briefly, leaving many questions unanswered and do not develop topics to any great depth.

If you are already a nursing manager or administrator or are planning to become a part of this challenging and important part of the nursing profession, you probably have discovered this problem; although there is a profusion of management applications available in the literature, they are scattered in a number of areas. And if you happen to stumble upon the application you need, it is not easy to relate it to a conceptual framework or to an overall philosophy of management. Thus you are left with a potpourri of articles and books that are as likely to be confusing as they are to be helpful.

The Management Anthology Series is designed to solve this problem by placing at your fingers a wealth of management information. Each book in the series focuses on one management topic; each is an anthology—a collection of the best selections from the literature—about a specific topic. The selections are chosen by talented people, usually nurses, who are experts in a particular field of management. These editors have generously added their own wisdom, opinions, and nursing examples to the management literature, producing not just a compendium of articles, but a logical conceptual flow of current thought by the most respected experts on that subject.

The selections in each book are chosen to provide a specific progression of concepts, and each book, in turn, contributes to the overall conceptual framework of the series. Each article in each book is an integral part of a set of beliefs, goals, and content.

PREMISE STATEMENT

This series of books on nursing management is based on several premises. The interrelated components of this belief system illustrate the dynamics of the world of nursing administration and can be seen as a conceptual framework that ties each theme and each book into an understandable whole. The five major components, or themes, are:

1. The organization, its people, and its structures
2. Management functions
3. Products, services, and cost effectiveness
4. Employee growth and satisfaction
5. Organizational security and longevity

Theme 1: The Organization, Its People, and Its Structure

The organization is the basic social matrix of the conceptual framework because it contains the concrete interpersonal and intergroup processes through which social action is accomplished. Formal organizations require conscious and purposeful cooperation among people. As individuals, we recognize that cooperation is crucial; through it we can accomplish purposes that alone we are biologically unable to do. Cooperation is essential to the survival of an organization, and because cooperation depends upon communication and interaction, organizational units are usually limited in size.

Maintaining the fabric and patterns of cooperative enterprise is the job of managers, who are responsible for organizational structures, lines of accountability and authority, and individual and group responses to change. Theme 1, the organization as the basic social matrix of our conceptual model, is shown in Exhibit 1.

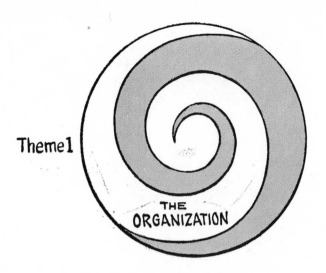

Theme 1

THE ORGANIZATION

Exhibit 1:   The Organization, Its People, and Its Structure. (Theme 1.)

Theme 2: Management Functions

When Og, the prehistoric caveman, and his fellow tribespeople realized that they had to produce food to fill their hungry stomachs, they started communicating, planning, organizing, assigning jobs (staffing), and counting the cost (control and feedback). They also looked to someone to lead them. They didn't realize it, but they had to engage in management functions in order to achieve their goal—full stomachs!

Throughout the ages, this has been true of all collective human endeavor. Whether we realize it or not, we must fulfill management functions if we wish to achieve our goals. Seen in this way, management is not something forced upon people, it is a set of processes that we create for ourselves in order to ensure that we will be productive, satisfied, and secure.

The group initiates management tasks: communicating, planning, organizing, staffing, controlling, and leading. These management functions are universal processes based upon a body of knowledge. When a group is small, the processes simple, and the products uncomplicated, each person may incorporate many elements of management into his or her day-to-day activities, such as patient care. As the group enlarges, as processes become more complex and specialized, and as products become harder to evaluate and count, the group designates certain people to do the jobs of management on behalf of the entire group.

In contemporary society, with its large corporations comprising thousands of people, various management functions have been assigned to certain people. This has led to a belief by some people

ix

that management is an unnecessary, arbitrary, and capricious group of people at the top. It is true in some situations that management has deteriorated to the level of capriciousness, but in the most it is, and must be, a way of group thinking, planning, acting, communicating, and influencing that makes life better for both workers and consumers. Management functions arising from the needs of the organization and its people are illustrated as Theme 2 in the developing conceptual model shown in Exhibit 2.

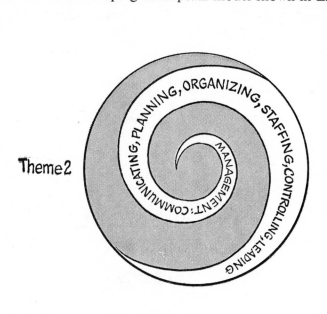

Exhibit 2. Management Functions. (Theme 2.)

Theme 3: Product and Service Cost Effectiveness

People band together as an organization to produce something. Nurses come together to provide a service called nursing care, which is the profession's most obvious product.

An enterprise survives only so long as its "official," publicly offered products or services are marketable, or valued by society. Society expects an organization to produce products or services of quality at prices that are justifiable. To achieve this, nurse managers are responsible for defining values, formulating criteria, devising measurement methods, and setting forth, in understandable terms, the quality and cost of nursing care services. Theme 3, product and service cost effectiveness, is shown in Exhibit 3 as one of nursing's outputs.

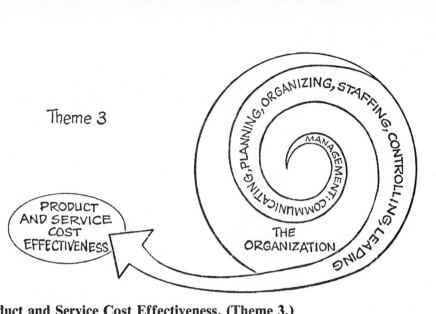

Exhibit 3.   **Product and Service Cost Effectiveness. (Theme 3.)**

Theme 4: Employee Growth and Satisfaction

People want to grow and develop, and work enterprise has the potential for being one of society's most powerful instruments for individual growth. People are always "wanting and growing"; when one need is satisfied, another appears. This process is unending, continuing throughout one's life. Therefore, nurse managers provide for: a motivating environment, job enrichment, and educational opportunities for the group's members. An important organizational output is personal and professional growth and the satisfaction of its members. This component of the developing conceptual model is shown in Exhibit 4 as Theme 4.

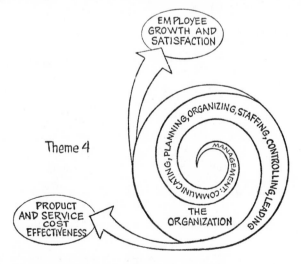

Exhibit 4.   **Employee Growth and Satisfaction. (Theme 4.)**

xi

Theme 5: Organizational Security and Longevity

Finally, organizations must engage in transactions with both the internal and external environments simply to survive, and even more importantly, to grow. If they cannot cope with their environments, they die.

Managers, who foster the quality of the group's transactions, develop sense organs to detect environmental changes. They forecast, plan, and develop strategies for survival and growth, always looking as far into the future as possible. The security and longevity of the organization is itself an organizational output. This is illustrated as Theme 5 in the completed conceptual model shown in Exhibit 5.

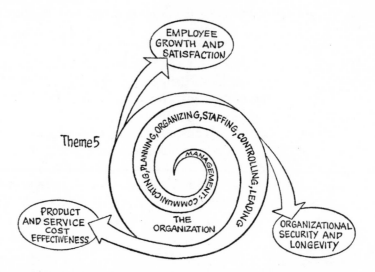

Exhibit 5.   Conceptual Model for the Management Anthology Series. (Theme 5.)

Summary

In summary, this conceptual framework incorporates the belief that a nursing organization must have three major outputs: cost-effective patient care; satisfaction and growth for its members; and organizational security and longevity. The absence or diminution of any one of these three can jeopardize the others.

The conceptual model illustrates the organization of the Management Anthology Series. The major principles of management practice are interrelated in a comprehensive conceptual framework, whose basic elements are derived from the theories and principles set forth by such authors as Drucker, Odiorne, Etzioni, Benne, Bennis, McGregor, Herzberg, and Argyris. The conceptual framework is divided into five major organizing themes, each of which has multiple subtopics that are the focus of one or more of the resource books. Some subtopics of each of the major themes are listed in Exhibit 6.

Exhibit 6.   Conceptual Framework for the Management Anthology Series.

Theme 1:   **The Organization** includes books on:
Organizational theories and structures
Organizational units and groups
Individuals in organizations
Intraorganizational conflict
Organizational change

Theme 2:   **Management Functions** includes books on:
Communicating
Planning
Organizing
Staffing
Controlling
Leadership in nursing

Theme 3:   **Product and Service Cost Effectiveness** includes books on:
Philosophy, goals, and criteria
Quality measurement
Maintaining cost effectiveness
Management goals

Theme 4:   **Employee Growth and Satisfaction** includes books on:
Motivation
Job enrichment
Development through education
Development of nurse managers
Personal growth

Theme 5: **Organizational Security and Longevity** includes books on:
Organization/environment relationships
Social policy and nursing services
The women's movement and nursing
New human services and technologies
Labor relations in nursing
Legal protection of patients and nurses
The politics of nursing

Of course, any conceptual framework represents an arbitrary, yet defensible, division of content. The purpose of the division is to simplify a universe of knowledge so that one can grasp its essential nature. This accomplished, one can then deal with the myriad of details that logically (and sometimes arbitrarily) follow.

The conceptual framework of this series of books is designed specifically to provide nursing administrators and managers with current, comprehensive, and practical resources for dealing with management problems and issues. Each book relates to just one facet of management, and each selection covers theory as well as practical applications for nursing management situations, making it possible to review current thinking and practice quickly and efficiently.

Theme Five: Organizational Security and Longevity

Our health care agencies are not closed systems. They are constantly interacting with outside influences that require changes in our old ways. Like it or not, our instincts to resist change—to hold on to the "knowns" of the past—must be sublimated if not overcome. As health care administrators we know that our eyes must be trained to see the elements that will influence and shape our organizational lives in the decades ahead. Many clues to the future are already around us in varying stages of development and clarity. The elements that will have an impact on nursing organizations lie primarily in technology, human values, and in economic and business trends.

Certainly health care is a technologically oriented industry. Not only are new generations of clinical technologies proliferating at an exponential rate, but also is the new world of automated information systems becoming an almost absolute must if health care professionals are to document, recall, and quickly utilize the rapidly expanding scientific knowledge base of patient care and its adjunct services.

Meanwhile, human values are changing at an unprecedented rate. Employees are asserting themselves. They are expressing themselves as persons with certain rights and with skills and knowledge that deserve respect and entitle them to input to management.

Economic and business trends in society at large are impinging upon the nurse administrator. Requirements for creative as well as sound fiscal planning abound. In the hospital world financial security is, to a large extent, dependent upon outside forces such as health systems agencies and reimbursement regulations.

The concept of cosmopolitanism is proffered by Leavitt and

others[1] as a way for an organization to keep up with and thus be ready for the future. Cosmopolitanism in this context refers to an organization's active encouragement of its members to have multiple affiliations—affiliations with and memberships in a wide variety of similar and "competing" groups. An employee with multiple memberships becomes an important source of information. But some organizations actively defend themselves against members who present too much disturbing information:

They don't want to be confronted with intelligence that disturbs their equilibrium . . . especially if the (new) reality requires them to change. For change is costly and disruptive . . . Professional people in organizations are particularly vulnerable to such rejection . . . Yet it is precisely by wandering abroad that organizations glean information that may be essential to their growth and survival. It is in this way that they identify new problems before they become critical and discover new opportunities before their competitors do. And in the process they may set up new and useful relationships with other groups [2].

In many respects, nursing is cosmopolitan by nature. The diversity of our areas of expertise and our work locales almost automatically creates a cosmopolitan attitude. On the other hand, concerns for the profession's future can inadvertently create a closed empire-building attitude that could result in professional and organizational provincialism.

To build and maintain strong and viable nursing organizations, nurse administrators must be open to news from the outside world so as to periodically redefine their mission. This means that we must engage in more give-and-take with employees, patients, taxpayers, governmental agencies, and the media. Nurse administrators must be cosmopolitan enough to seek real dialogues with their key publics in their environments.

The books in this theme, organizational security and longevity, deal with some of the obvious issues that will influence the viability of our health care organizations, that will mean their health or their demise. Organizational security and longevity are seen as a management-organizational output, as necessary, in itself as its services and growth and satisfaction of its employees.

This volume, edited by Dr. Janet Kraegel, contributes to theme five by showing how one can analyze operational-environment relationships as a major first step in becoming wisely cosmopolitan in one's environmental interactions.

References

1. Leavitt, H. J., Dill, W. R., and Eyring, H. B., *The Organizational World.* New York: Harcourt Brace Jovanovich, 1973, pp. 198, 199.
2. Leavitt, H. J., Dill, W. R., and Eyring, H. B., 1973, p.199.

Foreword

Our health care world today is turbulent compared with even fifteen years ago. Many nurse administrators recall the comparatively peaceful environments of days gone by—the days when the health care industry was less visible in the economic sphere and less regulated in the governmental arena.

Few would argue that this certainly is not the unchallenged, peaceful "sleeping giant" of a health care industry of fifteen years ago. It has finally been prodded quite enough. The giant is awake and is struggling to quickly learn how to move its gargantuan body nimbly and effectively among a crowd of other giants who have not been taking a nap. Our giant has awakened to a world of competition, regulation, diminishing resources, exploding technology, consumerism, diverse ideologies, civil rights and much more.

The nurse administrator is as exposed to these inner and outer environmental pressures as are the managers and executives of any large business or industry in the world today. The nursing sphere is buffeted, pushed, pulled, threatened and rewarded all at the same time for different reasons from different sources. Some threats and opportunities are predictable. Others come as surprises. The nurse executive and his or her organization must have keenly tuned sense organs to detect small changes in the atmosphere, obtaining clues that help to predict the larger "weather fronts" that are gathering force and strength and that will ultimately bring rain or shine. How much better it is to be warned and to be prepared so as to either take advantage or to shield oneself.

Thus, if the nurse executive's organization is to grow and prosper, he or she needs a model, a way of analyzing the agency and its environment, and must foster organizational sense organs and be willing to face the realities inherent in what is sensed.

In this book, the author, Dr. Janet Kraegel has gleaned some of the classic literature on the subject of organizational-environmental relationships. The reader is provided with scholarly, vital information that, when read carefully and thoughtfully, brings into focus one of the most elusive, yet crucial managerial skills: an ability to mentally step aside and to look, with comprehension, at one's organization—its transactions, strengths and vulnerabilities in its various environments. It is obvious that interorganizational transactions are necessary just to survive, but executives must attend to enhancing these transactions so as to foster growth. Survival is not enough. Growth is essential. Absence of growth is the presence of the beginnings of death. This anthology, by Dr. Kraegel, is one response to Maslow's challenge. He says that much more should be said and understood about the "relations between the enterprise and society, especially if we take into account the ways to keep the organization healthy over a period of a hundred years"[1].

Marlene Mayers, R.N., M.S., F.A.A.N.
Series Editor

References

1. Maslow, A. *Eupsychian Management: A Journal.* Homewood, Ill: Richard D. Irwin, Inc. and the Dorsey Press, 1977.

Table of Contents

Preface

As institutions gain in complexity, they increase their differentiation and change their relationships with the external social and cultural environment. They develop specialized "boundary-spanning" structures and more sophisticated means of communication, they establish interagency relationships to further their goals. These accommodations to the external environment are gradual but pervasive; they ask of organization administrators more than the intuitive responses which sufficed in a simpler day. This anthology presents a sampling of the theory which has been developed to explain organization-environment relationships; current literature is included to direct nurse administrators' attention to the strategies to be considered as their health care institutions move through these changes.

Until the 1960s most theorizing about organizations and management focused on the internal environment of the organization and emphasized goal achievement. Gradually this perspective changed as organizational theorists became increasingly sensitive to organizations' transactions with their external environment. Modern organizational theory is based on the premise that organizations influence and are influenced by their external environment considerably. The open-systems model is the theoretical tool used for conceptualizing the relationships between organizations and their environments. This volume focuses first on the immediate environment of the organization (the microenvironment) and then expands the focus to include more generalized environmental effects (the macroenvironment) and interagency relationships. Organizations' boundary-spanning structures, their transactions, and the strategies they use in their interagency relationships are described through the selections of this book.

Independent undifferentiated organizations respond to their problems by creating more procedures and more controls; in contrast, the answer to organizational problems for differentiated multisystem organizations is to build flexibility into the institution in order that it can respond to external demands and opportunities. Nurse administrators in the complex health care institutions of this era should be able to understand political organizational responses; they should be able to predict whether environmental forces will grow more or less favorable, and they should be able to assess their department's degree of interdependence with the consumer market and other groups of the external environment.

In Unit I the conceptual framework for understanding organization environment relationships is presented using a few key studies from organizational research and a description of the hospital as an open system. In Unit II are considered various components of an organization's external environment and their effects upon an organization's structure and performance. Examples from professional nursing literature are used for illustration. In Unit III, the responses of the organization to environmental pressures are emphasized; strategies discussed include cooperation, competition, negotiation, and planning of change. The intent of this anthology is not to offer specific administrative techniques; rather, the intent is to increase nurses' depth of understanding about 1) the context in which their organizations function, 2) the nature of their hospitals' political relationships, and 3) the range of responses that can be considered by nurse administrators who must adjust their departments to environmental forces that cannot be controlled.

Janet M. Kraegel, M.S.N., Ph.D.

Introduction

Nurse administrators are aware that events and processes which occur beyond the walls of their own hospitals can directly affect their ability to carry out their administrative functions. They realize, for instance, that a prolonged strike at a large community industrial firm could well be the cause of increased job applications, since inactive nurses married to workers in the industry are nudged into seeking an alternative source of family income. When the state changes its Medicaid reimbursement formula, the astute administrator may automatically request additional staffing positions in anticipation of an influx of patients newly eligible for coverage. Word of union activity in a hospital in a nearby community will increase the administrator's watchfulness for indications of similar behavior in his or her own institution. Social, economic, and political events which occur in the external environment of the hospital control the resources the hospital has available to it.

People frequently speak of things being beyond their control when they are referring to external environmental factors which affect what their organization is able to accomplish. They are referring to constraints or limitations placed upon their organization by the larger environment to which their organization delivers services.

This anthology identifies external environmental factors which influence organization effectiveness. It offers conceptual tools which practitioners can use for understanding the systemic nature of organization-environment problems.

Early studies of organization virtually ignored the constraints which the external environment places upon people's ability to achieve organizational goals. Until the 1960s management scientists concentrated upon the internal environment of the organization and the people working within. The organization was considered an instrument to be manipulated to the advantage of those who directed it towards the realization of ultimate goals. Only within the last two decades have scholars conceptualized organizations as open systems which are constantly being influenced by, and adjusting to, the external environment upon which they are dependent.

As organization scholars gained perspective through the open-systems conceptualization, there was increasing recognition that the true organizational objectives almost always lie outside the boundaries of the organization[1]; thus, measures of an organization's effectiveness can only be derived from an examination of the larger system, of which the organization is but a part.

This may be welcome news for the nurse administrator who has personally assumed the blame when his or her department failed to meet organizational objectives; on the other hand, the administrator who frequently assesses social, legal, economic, and political events occurring in his or her institution's external environment may be able to act to see that the department is prepared for the consequence of those events. Skilled nurse administrators who are aware of the significance of organization-environment relationships build a flexibility into their departments' plans. They realize that flexibility allows an organization to adapt to environmental contingencies constructively.

When administrators have a conceptual knowledge of how an organization relates to its larger environment, they are better able to recognize changing opportunities as they occur, and to plan accordingly. The conceptual framework most frequently used for understanding organization-environment relationships is modern organization theory. In Unit I several theoretical approaches to organizational theory are discussed, and Frank Baker gives an example of modern organization theory in the selection

entitled "The Changing Hospital Organizational System." Additional selections identify critical organization-environment interrelationships and describe the impact of social, political, and economic forces on organizations and their management.

An open-systems model of organizational effectiveness shifts from traditional focus on organizational goal attainment to a systems focus on the transaction processes which organizations use as they seek organizational goals. Organization-environment transaction relationships are considered the central ingredient in the modern definition of organization effectiveness[2]. Much of the recent literature adheres to this theoretical assumption.

Unit I selections are ordered so as to carry the reader sequentially through early open-systems organizational conceptualizations to show how scholars in the sixties built on the research of their colleagues. Eventually a fairly well formulated description of the organization's relationship to its external environment was developed, based on the general systems paradigm.

The term **environment** is a generic term which has many contextual definitions. Unit II describes how environment has been conceptualized in modern organization literature. Emphasis in Unit II is on the impact of environmental influences on the organization. Task environment, contingency theory, and broad social and political environmental considerations are the concepts discussed.

In Unit III the selections direct attention to the organization itself and its response to environmental influences.

Strategies managers use to accommodate the organization to environmental influences are categorized and illustrated. The selections throughout the anthology identify the critical factors nurse administrators should consider when planning the services of their departments. As managers, they are primarily responsible for the appropriate allocation of institutional resources. The people who administer institutions most frequently respond to external environmental influences by reallocating resources, assigning new responsibilities to personnel or creating new positions, acquiring new resources, or changing the nature of transactions between institutions and the environment. Nurse managers who function without an open-systems perspective frequently spend the better part of their time reacting to immediate problems as they are encountered. Nurses with a systems perspective, on the other hand, are geared to looking beyond the immediate problem to the larger context in which the problem has occurred. They attempt to comprehend patterns underlying the events and analyze the relationships between their organization and groups or individuals in the external environment. Such analysis opens up many more avenues of action for consideration as a way of alleviating the problem. The selection in Unit III by Harry Berg presents a simple example of the advantages of looking beyond the walls of the institution for problem solutions. Systems-oriented administrators solve problems by seeking to understand the true nature of the total system in which the immediate problem is but a part.

Open-systems theories of organization (frequently referred to as modern organizational theory) do not make it possible for administrators to predict the outcomes of their decisions accurately, although the confidence with which various scholars state their hypotheses would lead one to believe that this might be possible. The very nature of systems makes it difficult, if not impossible, to develop a definitive body of deductive theory about organization-environment relationships. Institutions are complex social systems in constant interaction with a changing environment. Under such conditions critical variables are uncontrollable, and a multitude of intervening variables intercede between the cause and the effect of a relationship. The complexity will not yield readily to our present state of the art in deductive research. As A. J. Melcher has observed[3], the formulation of scientific concepts to describe systems relationships, structure, and processes is difficult at best. Concepts must become measurable and defined in order to be useful. As yet we have limited understanding of the mediating processes at the various levels of organization systems. Even though the theoretical framework upon which this anthology is based is not highly developed, familiarizing yourself with this collection of research and applications should prove useful.

Glossary

Adapted and reprinted with permission from Warfield, J. N. and Hill, "Word Glossary," in Gordon, B.B. ed., *A Unified Systems Engineering Concept*. Columbus, Ohio: A Battelle Monograph, No. 1, June, 1972, Appendix A-4. Copyright 1972, Battelle Memorial Institute.

Adaptive system. A) A system that achieves insensitivity to environment or parameter perturbations by controlling its internal performance. B) A system that displays a persistence of success (or of acceptable performance) in the face of a changing environment and/or a deterioration of the performance of individual components.

Boundary. The minimum description required to distinguish a system from environment.

Closed system. A theoretical (formal) system that has no environment—hence the boundary is null.

Constraint. A restriction or limitation.

Control. That which, given a standard of comparison or means of verification, affords a means of directing performance in the direction of the standard of comparison.

Environment. One of two parts of the universe, the other being the system, each part being subject to arbitrary definition. Such definition is usually based on partitioning principles.

Analytical. A model of that portion of the universe not defined to be part of the system, usually limited to models of those components of the real world environment that are known to interact most strongly with the system.

Real world. Assuming the system to have been abstracted from the universe, that portion outside the system which can be observed to interact with the system.

Equifinality. A property of a system that it can reach the same final state from different initial conditions and in different ways.

Equilibrium. The property of remaining constant when multiple agents of change are present.

Feedback. The return of performance data to a point where they can be compared with objectives data, normally for the purpose of improving performance (goal-seeking feedback) but occasionally for the purpose of modifying the objectives (goal-changing feedback).

General systems. A field of study characterized by efforts to determine theoretical principles which can be applied usefully to systems of many kinds and from many disciplines, based on a belief that combining contributions from many fields will permit advances which can in turn be communicated back to specific fields.

Goal. Synonym: Objective.

5

Hierarchy. A partially-ordered structure of entities in which every entity but one is successor to at least one other entity, and every entity except the basic entity is a predecessor to at least one other entity. A hierarchy of systems represents a graded order of systems based on size. No superior-inferior relationship is implied. Each lower-order system is a subsystem of a higher-order system; each higher-order system contains a number of subsystems; each system can be related to a similar system at the same level.

Input to a system. That which cannot reach the system except by crossing the boundary from the environment to the system.

Interface. A common boundary between two or more items.

Open system. A system whose boundary changes with time.

Organismic system. An open system characterized by relatively fixed ordering of components, reproduction of the same type of species, and existence of a life cycle.

Organization. Strothers[4] suggests the following characteristics which apply to any aggregation of people described as an organization: A) Organizations are groups of two or more people B) in some kind of cooperative relationship to one another. C) This cooperation implies a collective goal(s) or output(s). D) The groups exhibit differentiation of function among members, and E) maintain a more or less stable and explicit hierarchial structure. F) Organizations exist in a total field—references to clients, milieu, inputs, equilibrium, and legitimation indicate recognition of this fact.

Output. That which passes from the system across the boundary to the environment.

Resource. Anything that can be made available that will be contributory to satisfying an objective.

Stability. The property of a system that makes it continue to remain in its present state regardless of changes in its environment.

Structure. The form or geometry of a system as opposed to its material content.

Subsystem. A defined part of a system.

System. A complex unity formed of many, often diverse, parts subject to a common plan or serving a common purpose.

Universe. The totality under consideration often separated into systems and environment.

REFERENCES AND NOTES

1. Yuchtman, E., and Seashore, S. A system resource approach to organizational effectiveness. *Am. Sociological Rev.*, 32, 1967.
2. Yuchtman, E., and Seashore, S. 1967.
3. Melcher, A. J. Promises, Problems, Realizations. In Melcher, A. J. (ed.) *General Systems and Organization Theory.* Kent, Ohio: Kent State University Press, 1975.

UNIT I   THE ORGANIZATION-ENVIRONMENT CONCEPT

- **CONCEPTUAL FRAMEWORKS OF ORGANIZATION THEORY**
- **THE GENERAL SYSTEMS PARADIGM**
- **ENVIRONMENT-ORGANIZATION DIMENSIONS**
- **APPLYING MODERN ORGANIZATION THEORY TO THE HOSPITAL SITUATION**
- **TRANSACTIONS AND RESOURCE ACQUISITION**

1. Conceptual Frameworks of Organization Theory

Managers in health care institutions face complex problems in their day-to-day functioning. Although they may recognize that events are caused by many forces working in complex relationship to each other, they frequently have difficulty understanding the pattern of the relationship and instituting actions which will result in desired outcomes.

Many conceptual frameworks for understanding organizational phenomena have been offered to the manager. The earliest ones, now called classical theories of organization, addressed the structures, processes, and role relationships within organizations. External environmental factors which affect an organization's goal attainment were not considered. Because of their failure to consider organizations as dynamic entities open to environmental forces, classical organization theories are said to deal with closed systems.

In the late 1950s, a new organization perspective was increasingly articulated as a corollary to the general systems theory of Bertalanffy[1], Boulding[2], Wiener[3], and others. Theorists began to conceptualize organizations as open dynamic systems, constantly shaping and being shaped by their environment. Organizational theory based on general systems theory is commonly referred to as modern organization theory. The systems model of organization emphasizes both the distinctiveness of the organization as an identifiable social entity and the interdependence of this entity with its external environment.

The most recent trend in organizational research is the incorporation of political considerations into organizational literature. Competition, negotiation, and bargaining are terms which increasingly occur in our everyday discussions as administrators. Today we are much more aware of the political context in which we function than we were even 10 years ago. Organizational theory and political theory are blending. This is particularly evident in the selections of the last half of this anthology.

As organization theory evolved from classical theory, it incorporated the human element, expanding into behavioral theory. Eventually consideration was given to the external environment as the open-systems paradigm became generally accepted.

CLASSICAL ORGANIZATIONAL THEORY[4]

Historically, the earliest students of organization were statesmen and philosophers chiefly concerned with the ethical base of political authority as embodied in platonic thought of the rational ideal, Thomas Hobbes' theory of social contract, and John Locke's doctrine concerning the consent of the governed.

Technological innovation and urbanization resulted in increased scale of organization and a concomitant need for social structuring to promote harmonious functioning within industrial organizations. In the early twentieth century, the point of view of social contract was extended specifically to industrial organizations through the writings of Mary Parker Follett[5]. The concept is still viable in organizational theory; i.e., no one should give orders, but both managers and workers should respond to the demands of the situation. The first scientific approach to organizations thought is credited to Frederick Taylor, who conceived of a theory of first-line management, task organization, and supervision for greater efficiency. The singular goal of Taylor's model, derived from economic theory, is profit maximization[6]. This rationalistic approach was adopted by Max Weber, who developed a theory of bureaucracy that is concerned with organizational rules of conduct and hierarchical control[7]. These "rational" theories of scientific management and bureaucratic control dominated early organizational research and continue to be a point of theoretical departure even today. The four major elements of this model are division of labor, scalar and functional processes, structure, and span of control. The organization is considered an instrument for ordering interpersonal relations toward the achievement of a prescribed goal. Function and resource considerations fit a master plan, outcomes are considered predictable, planning and control are emphasized, and the organization's environment creates a need for action[8].

THE BEHAVIORAL APPROACH TO ORGANIZATIONAL THEORY

The classical or rational model is useful, but it tends to obscure the human element and the internal purposes of the organization. As social scientists turned to examining the human factors, a second broad categorization of organizational theory evolved, commonly referred to as behavioral organizational theory. It presumes that an organization's behavior is the result of combined patterns of individual behavior; that is, the actors within the organization, rather than the organization itself, act. It further presumes that organizational behavioral processes must take into account the cognition, perceptions, beliefs, and knowledge of the actors; thus, rewards or goals are often complex[9]. The concept of the informal organization, as first identified by the Western Electric Hawthorne studies, is prominent in this body of research. Early contributors to this body of knowledge are Mayo[10] and Barnard[11].

MODERN ORGANIZATIONAL THEORY

A third broad categorization of organizational theory emerged in the 1950s. This approach attempts to integrate both organizational structure and behaviorism into the study of organization as an open system, and is generally referred to as modern organizational theory[12]. This research draws upon systems analysis and cybernetics to answer questions concerning mutually dependent variables which enable an organization to produce and survive. The open-system model of organization assumes that the behavioral factors and the physical environment are interdependent. They work together to achieve a balance which assures system integrity in the face of changing conditions[13]. The elements of organization are interrelated through information networks, control or feedback mechanisms, and decision-making processes.

Scott differentiates between modern organizational theory and general systems theory in terms of their generality. General systems theory is applicable to every system in the universe, whereas modern organizational theory focuses primarily upon human organizations. Baker[14] states that the shift in recent years from rational organizational theory to open-system concepts is in large measure related to the publication of Katz and Kahn's *The Social Psychology of Organizations*[15]. Other investigators include Georgopoulos[16], Yuchtman[17], Emery and Trist[18], and Thompson[19].

A subset of modern organizational theory known as "contingency theories" conceives of the relationship between the organization and its environment as being a contingent one; that is, the internal processes of the organization must be consistent with the requirements of the tasks to be performed, and there must be a fit between the internal organizational characteristics and external environmental requirements if the organization is to perform effectively[20]. Contingency theories are based on Bertalanffy's concept of equifinality, which postulates that an open system may reach the same final state from different initial conditions and in different ways[21].

The broad categorizations of organizational research just described indicate how research methodology has progressed from characterization of the bureaucratic phenomena to a search for explanations of and consequences of different structural arrangements and different environments. Efforts have turned from rationally finding the "one best way," to recognition that organizations have multiple goals, and that there are alternative ways to effectively attain them.

THE CONVERGENCE OF POLITICAL AND ORGANIZATIONAL THEORY

The twentieth-century press of technology for solutions to organizational problems diverted the mainstream of organizational thought from its initial ethical and political considerations[22]. Political scientists leaned heavily upon normative explanations and tended to use classical organizational theory only prescriptively. They use Weber's rational model in emphasizing public organization's bureaucratic features. Organizational theory is more likely to be used by sociologists studying political organizations than by political scientists, themselves. More recently, organizational scientists have been incorporating political concepts in their models[23]. Similarly, political scientists are drawing more liberally now upon modern organization concepts[24]. Actually, political scientists and organizational scientists encounter the same phenomena. Kaufman[25] suggests, however, that political theorists have been more willing to deal with intangible and nonrational aspects of human associations than the organization theorists because political outputs of social goals are difficult to measure. It is frequently useful for political and organizational scientists to use concepts general enough to encompass both forms of organization. Aldrich[26] suggests, for instance, that conceptualizing organizations as boundary-maintaining systems can provide explanations of power and authority in terms applicable to all disciplines.

In classical organizational theory, the organization is considered an instrument to be manipulated to the advantage of those who direct it towards the realization of an ultimate goal. The effectiveness of the organization is considered to be the degree to which it achieves its stated goal.

Selznick[27] was one of the first to provide empirical evidence contrary to this goal model. In his institutional analysis of the Ten-

nessee Valley Authority, he demonstrated that the authority was required to adapt to sizable external and internal environmental pressures in order to survive. The authority used self-defense mechanisms which resulted in structural transformation of the organization itself. Thus, Selznick suggests that an organization can be characterized as being a responsive institution rather than an instrument. These contrasting orientations to organization behavior are generally referred to as the closed system (goal model) and the open system (survival model). Etzioni[28] describes the open-system model as a multifunctional unit in which some means have to be devoted to such nongoal functions as maintenance, recruitment of means, and social integration. He states that the goal model is not the best possible frame of reference because of its unrealistic Utopian expectations. He goes so far as to suggest that public goals are not meant to be realized. If an organization were to invest means in public goals to such an extent that it served them effectively, its survival would be threatened and the organization would discard them.

Yuchtman[29] cites two unsolvable difficulties of goal attainment models: 1) there is not and cannot be agreement on the nature of goals; and 2) even if agreed upon, goals cannot be shown to be the properties of the organization itself. The alternative is a system model which emphasizes both the distinctiveness of the organization as an identifiable social structure and the interdependence of the organization with its environment. The organization-environment interdependence takes the form of transactions in which scarce and valued resources are exchanged. Yuchtman states that an excellent expression of the organization's overall effectiveness is the degree to which the organization is successful in acquiring scarce and valued resources. This is described in terms of the organization's "bargaining position." It can be argued that organizations are intimately linked to society's goals, which gives them legitimacy. Yuchtman argues, however, that the organization's contribution to the larger system must be regarded as an unavoidable and costly requirement rather than as a sign of success. On the other hand, the organization must produce some important output for society in order to receive, in return, some vital input. For this reason, the relations between the organization and its environment are a central ingredient in the definition of effectiveness.

The systems model is a powerful tool for conceptualizing the interdependence between the organization and its environment, and the input-output transactions that take place. Thompson[30], in general, substantiates these conclusions by saying that the open-system model shifts attention from goal attainment to survival and incorporates uncertainty by recognizing organizational in-

terdependence with environment. Though organizations exist with the consent of their environments, modern societies have yet to develop procedures for scientifically assessing the value of organizations to the societies they are developing[31].

Bogdan[32] studied a poverty program and arrived at the conclusion that there may be a whole class of organizations that do not fit into the theoretical model of organizations as instruments for attaining societal goals. He suggests that it might be useful to think of a large number of organizational types as expressive, rather than instrumental—they allow societies to act out their beliefs. Presenting an acceptable image legitimates the organization to the public and allows it to operate freely. He suggests that the delicate relationship of goals, as symbols of consensus and instruments of power, can direct us to an area of organizational theory that needs more exploration; namely, the relationship between conflict and the functionalistic models.

In summary, two conclusions which may be drawn from modern organizational theory are the following: 1) effectiveness criteria must reflect the entire input-process-output cycle; and 2) effectiveness criteria must reflect the interrelationships between the organization and the larger environment upon which the organization depends[33].

2. The General Systems Paradigm

The systems perspective is promulgated in this volume because there appears to be no better way to understand organization-environment relationships. Problems in environment-organization relationships can best be understood by defining the functions that the organization is to perform; the significant relationships that exist between the people, resources, and processes of the organization; and the collection of interacting organizations, groups, and persons in the external environment. Because organizations can seldom attain their goals independently, they must establish exchange relationships with other organizations.

If one takes hospitals as an example, the functions they are to perform is to provide health services, educate health care workers, and continue to develop clinical knowledge for health care practice. The hospital is directly dependent upon the environment for its inputs of patients, physicians, other personnel, technology, supplies, and capital. Groups and organizations with whom the hospital usually interacts include governmental agencies, voluntary health agencies, professional health associations, philanthropic foundations, financial institutions, educational institutions, vendors, and other hospitals. In addition, the characteristics and needs of the community in which the hospital operates shape the structure, processes, and output of the hospital. The age, occupation, and per capita income of community residents affect the utilization patterns of the hospital, the personnel the hospital is able to obtain, and the capital resources directly available. It makes a difference whether the other hospitals in the area are highly competitive or cooperative, whether the physicians are highly dependent upon one hospital or have many to choose from, whether the community is conservative or liberal, whether the unions are strong or weak, and whether the setting is rural or urban.

15

The jump from highly abstract general systems concepts to pragmatic considerations such as the foregoing is a challenge scholars are constantly addressing. The application of systems theory to complex organization phenomena is difficult to establish empirically. Bertalanffy states, "There is a great and perhaps multiplicity of approaches and trends in General Systems Theory which is uncomfortable to those who want a neat formalism. Future developments will undoubtedly lead to future unification."[34]

There is considerable misunderstanding concerning what general systems theory is and what it is not. In the organization theory literature of the 1960s, there was considerable controversy over the systems approach. The scientific community had several bouts of indigestion before it could begin to assimilate the new paradigm. Acceptance of the open-systems paradigm for the study of organizations is not universal even today. Concern with intra-organizational phenomena or the "closed-system" orientation is still common among many organizational researchers. Evan attributes the persistence of traditional closed systems thinking to a microlevel perspective as follows:

a microlevel perspective (i.e., a concern with intraorganizational phenomena involving reliance on individual, interpersonal, group, or subunit levels of analysis) is still common among many organizational researchers. This, in turn, encourages a closed-systems orientation to organizational phenomena in spite of the attention paid in recent years to open systems theory. Associated with this view of organizations are two contradictory assumptions that have rarely been articulated: (1) that an organization—indeed, any type of organization—is a microcosm of the larger society; and (2) that an organization is a type of social system with a substantial measure of functional autonomy from its environment.

The microcosm assumption appears to be persuasive if we consider, for example, such ubiquitous problems as the quality of leadership, the rationality of the decision-making process, and the degree of inequality in reward systems within an organization as compared with the surrounding community or the society as a whole. By studying the internal structure and functioning of an organization, the organizational researcher is in effect studying the larger and more complex phenomena of the environing society.

By contrast, the functional autonomy assumption is conducive to a conception of organizations as having generic properties, for example, hierarchical, formal attributes of structure and a tendency toward internal differentiation for the purpose of securing coordination and control. Whatever their purported objectives and whatever their interrelationships with their surroundings, organizations are conceived as having properties that transcend the culture and the social structure of the society in which they are embedded. Hence, they may be studied as if they are functionally autonomous.

Paradoxically, both assumptions, although mutually exclusive, encourage inattention to the encompassing social structure as it relates to organizations. Notwithstanding these assumptions, organization theory has recently witnessed a marked interest in going beyond the boundaries of organizations for the purpose of studying interorganizational relations. If this trend persists and becomes more per-

*vasive in the future—and there is every reason to believe that it will—a prediction **I** made a decade ago may yet be borne out: "Systematic inquiry into the interactions among various types of organizations may not only unearth new intraorganizational phenomena and processes, but may also provide the wherewithal for bridging the gap between the microscopic **organizational** and the macroscopic **institutional** levels of analysis."[35]*

3. Environment-Organization Dimensions

In 1966, Stogdill wrote that "systems is coming to be recognized in the social sciences as a respectable activity"[36]. At the time, he undertook the identification of the major dimensions of organization to assist students confronted with numerous fragments of modern organization theory. His goal was to account for the most critical problems in organization relationships.

A wide range of organizational relationships is identified by Stogdill's three-dimensional panel: 1) the classical theory and interbehavioral segment, 2) the personal-organizational relations segment, and 3) the organizational-environment relations segment. Because the third segment deals with factors outside the organization which affect the organization's ability to cope with environmental stress, it is excerpted here. (All three segments are considered to be part of a single coherent system but space limitations prevent Stogdill's article from being reproduced in its entirety.)

The term "survival" is prominent in Stogdill's description of environmental-organizational relationships. During the sixties organizational theorists commonly conceived of organizations as organismic systems, that is, alive like animal or plant life. Kast and Rosenzweig challenge the transposition of organismic systems concept to social organization in their article reprinted later in this unit. Organismic systems terms such as entropy, negentrophy, equilibrium, steady state, and survival rarely appear in organization literature today.

DIMENSIONS OF ORGANIZATION THEORY

By Ralph M. Stogdill

Reprinted from *Approaches to Organizational Design* edited by James D. Thompson, by permission of the University of Pittsburgh Press. © 1966 by the University of Pittsburgh Press.

THE ENVIRONMENTAL-ORGANIZATIONAL SEGMENT

An organization is in part a product of its physical and cultural environment. The physical environment and the nature of the resources available may place constraints upon the kinds of activities in which the organization can engage. The societal environment may prescribe the aims and structure of organization, as well as the right to organize.

An organization engages in an exchange with its environment. The physical media of exchange will be determined in part by the resources and materials provided by the environment and in part by the social value placed on the available materials by the members of the larger society.

The viability of an organization is firmly rooted in the relationships that it maintains with its environment. The survival of utilitarian organizations depends upon their ability to extract from the physical environment those materials necessary to sustain their operations. Unless an organization gains unusual power, its aims and activities must be such as are tolerated by the larger social system of which it is a part. In order to survive crises an organization must maintain the internal mechanisms necessary for coping with change.

The three panels of variables (external constraints, exchange with environment, and survival capacity) shown in Exhibit 1-1 do not account in an exhaustive fashion for the possible relationships between an organization and its environment. But they appear to account for the most critical problems involved in those relationships.

EXTERNAL CONSTRAINTS

The environment imposes numerous restraints upon human organization. Despite man's inventiveness in exploiting his environment, climate and material resources have controlled the development of human societies to a marked degree. In Arctic lands only small, loosely organized bands have been able to find the food, clothing, and materials necessary for survival. Temperate lands provide enough food and materials to support large, complex, and densely populated societies. It is observed, of course, that societies differ in their utilization and exploitation of the same available resources. The kinds of organizations that develop in a society are determined to a very large degree

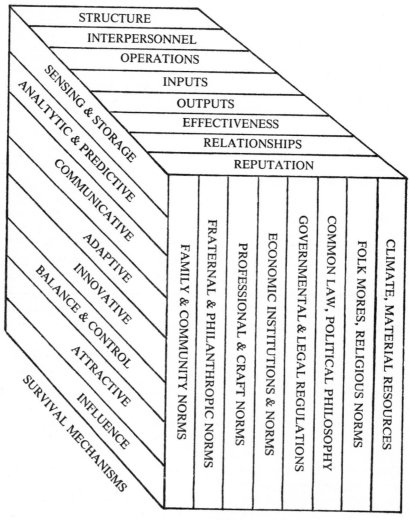

EXCHANGE WITH ENVIRONMENT

STRUCTURE
INTERPERSONNEL
OPERATIONS
INPUTS
OUTPUTS
EFFECTIVENESS
RELATIONSHIPS
REPUTATION

SENSING & STORAGE
ANALTYTIC & PREDICTIVE
COMMUNICATIVE
ADAPTIVE
INNOVATIVE
BALANCE & CONTROL
ATTRACTIVE
INFLUENCE
SURVIVAL MECHANISMS

FAMILY & COMMUNITY NORMS
FRATERNAL & PHILANTHROPIC NORMS
PROFESSIONAL & CRAFT NORMS
ECONOMIC INSTITUTIONS & NORMS
GOVERNMENTAL & LEGAL REGULATIONS
COMMON LAW, POLITICAL PHILOSOPHY
FOLK MORES, RELIGIOUS NORMS
CLIMATE, MATERIAL RESOURCES

EXTERNAL CONSTRAINTS

Exhibit 1-1.   The Environmental-Organizational Segment

by the technologies utilized for exploiting the environment. Udy[37] in a study of 150 non-industrial organizations found that tillage, construction, animal husbandry, and manufacturing tend to be carried out by permanent organizations; and hunting, fishing, and collection by temporary organizations. Tillage and construction tend to be more complex in task structure than hunting, fishing, and collection. Organizations carrying on complex processes tend to be more highly structured in regard to status levels than organizations carrying on simpler processes.

The social environment must be regarded as a far more potent force than the physical environment in determining the purpose and form of organization. Religious belief has played a major part in the development of large, complex, stable societies. The concepts of deity and moral principle have provided relatively stable standards for the legitimation of societal authority structures. The institutions and organizations of a society derive their rights and obligations from the still higher authority defined by the folk norms and religious doctrines of the dominant members of the society. Religious institutions have tended to support and stabilize governmental and economic organizations, as well as family structure. The reinforcement of religious belief at the family level has tended in turn to establish a wide base of legitimation among the individual citizens of the society[38].

Even when rulers have succeeded in imposing a new religion upon conquered countries, they have seldom been able to displace the folk norms of the people. Governments fall, and one religion replaces another, but the folk norms and common law tend to endure. Their stability and power are such that as a social system becomes stabilized they can often be appealed to above the codified law of the land. They too provide a base for legitimation and support of the institutions and organizations of a social system and at the same time restrict their rights and privileges.

In recent centuries, particularly in Western cultures, religious doctrine has yielded considerably as the ultimate source of organizational and governmental authority. Common law and political philosophy have been utilized increasingly as bases or criteria for the legitimation of organization. The appeal of political doctrine appears to derive from the fact that it broadened the base of ultimate authority and formulated the rights of individual man as founded in folk norms and common law. The right of individuals to organize for the accomplishment of a common purpose that is sanctioned by the societal norms has provided an instrumentality for the economic, educational, artistic, and cultural enrichment and advance of societies that maintain the right.

Governmental and legal regulations impose restraints on organizations in all societies with centralized governments. Udy[39] has shown that "societies with centralized governments are more likely to possess complex hierarchies of general social stratification than are societies without centralized government." The stratification of a social system and of its component organizations provides an effective mechanism for the imposition of restrictions defined by laws and governmental regulations. Thus, the right of an organization to pursue its purposes is subject to some restraint in most societies. The same normative system that legitimizes the right to organize also legitimizes the imposition of restraints designed to preserve the social system as a whole.

The economic philosophy, norms, and institutions of a society determine in part the purpose, form, and structure of the organizations it will develop. Udy[40] has shown that the ownership, structure, recruitment practices, and reward systems of organizations are closely related to the major kinds of economic activity that support a society.

Most societies develop professional specialists who have influence and often authority by virtue of their specialized knowledge and skills. If the specialists are numerous enough to form a professional society, the society is likely to formulate norms and standards that regulate the conduct of its members and define the permissible relationships between specialist and client. Thus, the professional society and the trade association serve to protect their members in the maintenance of high standards of technical performance. Barnes[41] reports that engineers in an industrial firm who identify themselves with the norms of their profession tend to be more productive and to present greater problems to management than do those who identify with the norms of the firm. Blau and Scott[42] report similar findings.

The norms of the fraternal and philanthropic organizations of a society may permeate and influence the conduct of all organizations in the society. Family norms and local norms may also influence the practices and activities of the organizations in a community.

An organization is subject to great variety of constraints. Environmental factors, both physical and social, operate to delimit the aims and activities of organizations. But most of the constraining factors can also be converted by a resourceful organization into rights and opportunities for growth and achievement.

EXCHANGE WITH ENVIRONMENT

An organization may be regarded as an agent of exchange with its environment. The exchange may be of minimal character.

However, governmental, economic, educational, and religious organization tend to make major impacts upon the social system of which they are a part[43].

All interbehavioral variables constitute avenues of exchange between an organization and the larger social order. Contacts and negotiations between organizations tend to be carried out by members with similar functions and equivalent status rankings. The conduct of business is made possible by the fact that for almost innumerable kinds of transactions, single representatives are empowered to commit their respective organizations to binding agreements. Thus, organization structure may be regarded as providing an avenue of exchange.

The interpersonnel variables provide several channels for exchange between the individual and the organization, as well as between the community and the organization. Mutual expectations between individuals and the organization are communicated and reinforced through interaction, intercommunication, and interpersonal affect. A member's identification with norms and reference groups in the community may affect his relationships with the organization, thereby facilitating or hindering effective exchange.

The operations of an organization provide the most easily observed channel of exchange between an organization and its environment. Many of the inputs of an organization are provided by the members of the larger society. The resulting outputs have some positive or negative social significance, even if minor in nature. If an organization operates on material objects, they must be extracted from the environment and may change it to a marked degree. The manufactured product, in turn, makes a major impact upon the society that utilizes it. Thus the input and output variables also constitute dimensions of exchange.

The relationships that an organization maintains with the community and with other organizations and institutions involve exchange of social values. Its reputation or image also may be regarded as involving an exchange of values. Riley and associates[44] suggest that an organization has many publics and that its effectiveness in serving them may be determined by the differing images that it presents to its publics.

The exchange relationships between organizations and environments may change not only the social system of which they are a part, but even the surface of the earth that they inhabit.

SURVIVAL MECHANISMS

Often an organization is developed to attain a single specific objective. Once the mission is accomplished, the organization is disbanded. But many kinds of organizations have continuing objectives that extend into the indefinite future. Their failure might create hardships for large numbers of persons and in some cases for the society as a whole. Thus, many organizations regard survival as a fundamental objective, although it is not stated explicitly as such.

The survival of an organization is rooted in the relationships that it maintains with its physical-social environment. It must be capable of coping with environmental change. In other words, an organization should provide itself with the functions, arrangements, and mechanisms necessary to assess its internal condition, as well as those needed to assess the presence and trend of environmental change[45].

There is a rather large body of literature on the problem of organization change. The principles of change are discussed by Ginzberg and Reilley[46]. Case studies are presented by Ronken and Lawrence[47] and Jaques[48]. The readings edited by Lawrence and associates[49] and Bennis, Benne, and Chin[50], cover case histories, theory and research. Baum[51], Blau[52], and Gouldner [53] discuss bureaucratic structures that are resistant to change.

Organizations often appear more sensitive to the impact of external change than to the necessity for internal change. But this is not always true. The nature of organizational mechanisms for sensing internal and external change is not well understood. Organization analysts usually recommend the maintenance and encouragement of upward communication as a means of sensing the need for internal change. There can be no doubt of the value of this recommendation. But members in the lower and middle echelons of a large organization are not always aware of the exact nature and magnitude of internal problems. Top management seems to need to be better sensitized to the importance and operation of basic dimensions of organization. Problems arising in connection with variables that are not regarded as an integral and legitimate aspect of organization are likely to be ignored or rejected. This observation applies in particular to interpersonnel variables.

The memory of an organization resides not only in the memories of its individual workers, but also in its repository of written records, when such are available. Ready access to information in storage often provides an important survival advantage. But memory may constitute a handicap as well as an advantage. Organizations appear particularly unable to forget unfortunate incidents and unfavorable outcomes. Such memories of unpleasant events often lead to the rejection of proposals for change and innovation without adequate consideration of their possible advantages.

Analysis, diagnosis, and evaluation are facilitated by a sound body of professional knowledge. Better methods are available for determining the need for change in structure and operations than for change in personnel. Changes in structure are often recommended as cures for problems in interpersonnel. Systematic and comprehensive research is needed to isolate the conditions that determine and limit various problems in interpersonnel as well as the relation of such problems to structure and operations. Remedial action should be based on diagnosis that considers the possible effects of, and the new problems likely to be created by, the action.

Certain kinds of organizations maintain such delicately balanced relationships with their environment that their survival depends upon the ability to predict change. Military organizations maintain departments of intelligence that are responsible for collecting and evaluating information that will be useful for predicting enemy action. The sales departments of large industrial and retailing organizations maintain departments whose function it is to anticipate consumer demand and ability to buy. It is costly and hazardous to manufacture or store more than can be sold. The maintenance of an effective arrangement for predicting change and trend is coming to be recognized as a prime necessity in various kinds of organizations.

The importance of communications has been discussed in connection with the problem of internal sensing. Communication provides a primary channel for sensing external conditions and for effecting an impact upon the social environment. All members of an organization constitute potential or active channels of communication between the organization and the community. Members of the lower echelons do not always communicate what members in the higher echelons would like to have communicated. Unofficial communications are often more potent than official ones in determining the public image of an organization.

Although numerous organizations maintain specialized functions or departments that are responsible for sensing, evaluating, and predicting change, few maintain specialized functions for adapting to change. The intelligence function may collect and evaluate information regarding the nature of change. The planning function may anticipate change and develop plans for coping with it. But these functions are seldom responsible for the initiation of action designed to cope with change. Such action is usually initiated in higher echelons, which are responsible for policy formation and coordination of activities. The physical separation of the executive function from sensing, predictive, and evaluative functions renders an organization vulnerable to breakdowns in communications. This vulnerability resides not only in the failure to communicate, but also in the frequent reluctance of higher echelons to accept as authentic the communications received from lower echelons. Lower echelons also exhibit reluctance to accept communications that involve changes in role performance. Once the structure and operations of an organization have become stabilized, all status levels tend to resist change. Research results suggest that the acceptance of change is facilitated when all members who will be affected are involved in the task of planning for change.

With rapid advances in technology, the survival of an increasingly large percentage of organizations depends upon the capacity for innovation. Product innovation often involves innovation in organization structure and operations. Organizations tend to be conservative. Changes in structure and operations always involve a cost to the members in that they must redefine their roles and change their ways of doing things. Members become accustomed to the fact that the coordinative requirements of the organization set bounds to the initiative and performance of each role. Innovation and creativity require that certain roles be defined with wider areas of freedom than is customary for routine operations.

One of the triumphs of the classical (organization) theory has been the development of principles and procedures for coordinating and controlling the operations of organization. Poor coordination in industrial organizations can result in bankruptcy. It would be foolishly costly, for example, to permit most departments to manufacture materials for 1,000 assemblies if the department responsible for one critical part could make only enough to complete 100 assemblies. As the operations of an organization become more complex and as its relationships with its environment become more sensitive, the demands for coordination tend to increase.

Organization control is a function designed to determine the extent to which performance corresponds with stated objectives. Accounting reports, productivity charts, progress reports, and related oral and written communications enable members in coordinative positions to assess the extent to which an organization is performing as it is expected to perform. The necessity for control increases as the margin between success and failure narrows. The higher the level at which first-line operations are coordinated, the greater the need for prompt and accurate control procedures. Immediately and constantly available control information is needed for the effective operation of the automated plant.

Academic training and experience in operative positions of organization do not necessarily provide an adequate basis for understanding the extent to which coordination and control influence the definition of roles. Unsatisfactory role definition is usually attributed to the whim of supervisors or managers. This is often but not always the case. The greater the demands for organizational coordination and control, the more closely must role performance be defined, regardless of the personality or inclinations of supervisors. This fact has induced a number of large corporations to disperse their manufacturing operations. Each of several subunits can usually maintain a higher degree of freedom for initiative in role enactment for the attainment of a single subordinate objective than is possible in a larger organization that must coordinate operations for several subordinate objectives. Formalized control procedures are associated with complexity of operations. They are also associated with the depersonalization of performance.

Organizations differ in the mechanisms they devise for attracting members and customers. The purpose of an organization may be sufficient in itself to attract a closely defined category of members. Usually an organization must render a minimally satisfying service in order to attract members or customers. Many organizations depend primarily upon their reward (salary) systems to attract and hold members. But an adequate pay schedule does not necessarily guarantee the loyal support of members. Many other factors, such as recognition of performance, opportunity for contribution and advancement, security of position, and satisfaction of expectations, tend to strengthen member support of the organization. Advertising and a reputation for providing a satisfying product or service tend to attract customers.

As a society tends increasingly to grant legally defined privileges to organizations and to regulate their activities on a differential basis, they tend to develop mechanisms for influencing legislation and public opinion. Public relations departments and lobbyists have a specific function of influencing the environment to the advantage of the organization.

4. Applying Modern Organization Theory to the Hospital Situation

A number of problems are encountered in applying systems concepts to organizations and their environmental relationships. One problem is the fact that complex organizations are poor self-regulators, especially in terms of longer-term goals and global viability. Child[54] argues, for instance, that managers of some organizations may have the opportunity to control contingent factors or even ignore their implications for a considerable period of time. It is difficult to provide empirical evidence to identify cause and effect relations when the vagaries of human behavior have such a strong effect on organization behavior.

A second problem in systems theory application is the diffuseness of definitions of systems concepts and the nearly complete absence of ways to measure many of the concepts, such as boundaries of a system, the degree of organization, linkages, or the probability of interaction. Researchers have limited success in operationalizing these terms.

A third problem is the fact that open-systems approaches must allow for varying numbers of factors or influencing forces beyond the immediate ones of the problem being considered. This makes it difficult to control or systematically to account for variance in the variables being studied. A focus on the dynamics of interrelationships demands a different research strategy than the linear models traditionally utilized. The following selection by Baker[55] illustrates one way to approach the research problems. An open-systems model is used in the study of the organizational metamorphosis of a mental hospital to a community mental health center. Though this early model is largely conceptual, it offers the first step in applied research; namely, the formulation of the right questions to ask. The selection carries the reader from the more theoretical conceptualizations of Stogdill to the dynamics of actual hospital-

29

community relationships. The situations encountered as the mental hospital relates to the community are similar to those encountered by any health care institution that tries to adapt to the unique needs of the people in the community it serves.

THE CHANGING HOSPITAL ORGANIZATIONAL SYSTEM: A MODEL FOR EVALUATION

By Frank Baker

Reprinted with permission from *Man in Systems*, edited by M. D. Rubin. Published by Gordon and Breach Science Publishers, Inc., 1971. Preparation of this paper was supported by NIMH grant NH-09214.

As the concepts of community mental health gain wider acceptance, many mental hospitals are attempting to take central roles in newly developing community-oriented health programs and changes are being attempted in the organization and services of these institutions. This paper describes an open-systems organizational model that is being employed as a conceptual framework in the study of Boston State Hospital as it undergoes an evolutionary metamorphosis from state mental hospital to community mental health center[56].

In attempting to evaluate the functioning of a large mental hospital within a comprehensive community mental health program, the usual problems of adequate program evaluation are compounded. In most cases the demonstration of effectiveness in such outcome measures as length of hospital stay, discharge rates, and readmissions is dependent upon not only intentional variation in one program variable but upon multiple variables interacting with each other in interdependent ways. In addition to developing indicators of certain outcome characteristics, it is particularly important to study the process by which the organization searches for, adapts to, and deals with its changing goals.

Social science research on mental hospitals has focused principally on **intraorganizational** process with an implicit assumption that the organizational problems of the hospital can be analyzed by reference exclusively to its internal structure and functioning. The classical organization models employed in this type of research have been based for the most part on this closed-systems approach.

The community mental health center idea emphasizes the importance of increasing interaction and interdependence between the hospital, the community, and other care-giving organizations, with an increasing interpenetration of traditional boundary conditions. For example, since one of the primary aims of the community mental health center is continuity of care, permeabilility is required not only between the various subparts of the center but also between it and other agencies in the larger care-giving network.

In the evaluation of the functioning of the hospital as a community mental health center, it is necessary to look beyond the

31

hospital-based program to the community's total mental health network. One must consider **extraorganizational** and **interorganizational** processes as well. Three areas for focus in this type of evaluation are:

1. The intraorganizational processes of the changing state mental hospital

2. The exchanges and transactions between the changing hospital and its environment

3. The processes and structures through which parts of the environment are related to one another

OPEN-SYSTEMS ORGANIZATIONAL THEORY

The concept of the "organizational system" is a commonplace in organization theory today, and Scott[57] argues that modern organization theory differs primarily from classical and neoclassical organization theory in its acceptance of the premise that the only meaningful way to study organization is to study it as a system. Katz and Kahn[58] have recently written a book which takes an open-system approach to the study of the social psychology of organizations. They attempt to spell out open-systems concepts as a framework for organizing the research on complex organizations, even though much of that research was generated from either a closed-systems approach or a fragmented or nonsystems approach to theory.

The importance of the environment and of interorganizational relations in the conceptualization of the functioning of an enterprise has become the subject of increasing attention by organization researchers and theorists[59].

In England, workers at the Tavistock Institute of Human Relations have been active in developing organizational systems concepts, much influenced by Bertalanffy's work[60],[61] which first revealed the importance of a system being open or closed to the environment.

The open-system model of the Tavistock group assumes that 1) organizations are defined by their primary task; 2) organizations are open systems, i.e., an organization admits inputs from the environment, converts them, and send outputs back into the environment; and 3) organizations encounter boundary conditions which rapidly change the characteristics of the organization. The conceptual model for research on the mental hospital in transition that will be presented here is based on these same assumptions.

PRIMARY TASK

In order to analyze the internal activities of a mental hospital, it is necessary to consider the organization's directed impetus toward some goal in relation to its community. The systems model is often contrasted with the more traditional "goal model." However, the open-systems model is neither divorced from the goal approach nor is it in opposition to the goal approach. In order to survive, the organizational system must fulfill the function of achieving goals which define overall objectives and it must further develop sets of subgoals to be accomplished by subsystems such as roles and departments. The distinction between the two models for the researcher is that the systems model points out the importance of doing not only final-outcome analysis but also process-oriented analysis focusing on the intermediate outputs of a system.

Rice first introduced the concept of "primary task" to discriminate between the varied goals of industrial enterprises. He defined the primary task as the task that an institution had been created to perform. In a later book, Rice[62] recognized the difficulty of treating the organization as if it had a single goal or task, and he cited the teaching hospital, the prison, and mental health services in particular as examples of institutions which carry out many tasks at the same time.

Like most complex formal organizations the mental hospital has multiple goals and performs many tasks simultaneously. A mental hospital admits patients, provides therapy and custodial care for patients, attempts to place patients in the community, provides employment for mental health workers, looks after its employees, conducts research, keeps records and accounts, and if it has educational programs, provides professional training for nurses and residents. Usually, there is no settled priority of goals for an organization and any one of the goals becomes primary in the hierarchy at a given time according to the balance of forces then operating.

Multiple goals are not necessarily compatible and involve competition for scarce resources by those subparts of the organization which are more committed to one goal over others. For example, the organization that best fits the task of training residents is not necessarily the same one that best fits the task of returning patients to the community as quickly as possible. Sofer[63] has shown that the organizational structure required for research programs in a mental hospital does not easily fit with the rest of the hospital.

Difficulties may arise in defining the one primary mission or task of an organization because certain goals are denied and consequent ambiguities may confound the researcher's choice of appropriate criteria for judging task performance. Traditional studies of organizational effectiveness based on success in goal achievement have depended too much on what organizational spokesmen have said. As Etzioni has pointed out, the goals which an organization claims to pursue (public goals), "fail to be realized . . . because . . . they are not meant to be realized"[64]. It is too easy to confuse the convenient public fictions of official pronouncements with the actual functioning of a system. The stated purposes of an organization in its charter or other official papers or the reports of leaders may give a very distorted picture of the functions of the enterprise. Changes in organizational objectives as stated by officials do, nevertheless, provide a good beginning point for finding out about changes in the system which have taken place.

One of the ways the environment affects the setting of goal priorities in the mental hospital is through ideological penetration— the values which come in from the outside. Mental health facilities considered as organizations grew out of a complex interaction of cultural and technological systems. Cultural values or belief systems determine within broad limits what the goals of organizations will be. Currently, a new belief system is developing among mental health professionals which has been labeled the community mental health ideology.

As the community mental health ideology penetrates a hospital, it can affect the priority of goals set for the hospital. For example, the ideas of community mental health imply extension of the hospital's goals of patient care and treatment to include improvement of the mental health of a population through increasing attention to programs of primary and secondary prevention. These new goals may produce changes in the task hierarchy of a hospital. A community mental health ideology scale has been developed which is being employed in the study of the relation of this attitude to a program development[65].

INPUTS

Basic to the conception of an organization as an open system is the assumption that any enterprise, or parts of it, can be characterized as admitting inputs from the environment, converting or transforming them, and sending outputs back into the environment. The inputs to an organizational system include people, values, economic resources, physical facilities and technology—the variables which, as they are operated on in various ways, determine the outputs. The subsystems variables, the internal structures, attitudes, and activities account for the conversion or transformation of the inputs into outputs. The result of the internal processing is a set of output variables which are usually used in defining the effectiveness of the system. These outputs from one system in turn become inputs for other systems. These three sets of variables constitute the very core of the open-system organization theory.

The major inputs to a community mental hospital consist of people. The hospital takes in those defined as mentally ill and it is these patients that correspond to the "raw material" which is processed in an industrial enterprise. Just as the qualitative and quantitative aspects of the raw material available from the environment to an industrial system are major constraints on its productivity, so it is with the characteristics of patients. The patient input to the mental hospital may be examined in terms of demographic characteristics such as age, sex, race, socioeconomic status, religion, and education. For example, a large number of geriatric chronic patients provide a major constraint on what the hospital can accomplish in terms of treating its patients successfully and returning them to the community. The prehospital level of adjustment of the patient input, i.e., how sick the people are, is another constraint. Thus it is apparent that the nature of the population served by the hospital is extremely important in determining the functioning of the system. As a state hospital takes responsibility for specific geographic catchment areas with resulting changes in the input of patients to the

system, the internal structure and functions of the hospital and the character of its output will be affected. As an increasing number of individuals are seen on an outpatient basis, the balance between inpatient and outpatient programs is bound to be altered.

The rate and types of patient input are also determined by hospital-community relations. The attitudes of the community served by the mental health center will determine its use of the center. This includes both the attitudes of the sick, their families and friends, and also the attitudes of the other caregiving agencies who refer patients to the hospital. The analysis of the input of patients can be approached by analyzing the paths or routes through which patients enter, are treated, and leave the hospital, and by attempting to describe and assess those various paths.

Another kind of human input to be considered is the personnel coming into the hospital system, including both the fully professionally trained as well as the residents, student nurses, and others who receive professional training. The level of and type of professional skills, the personality, attitudes, interests, goals, ideologies, and habits of the personnel are important input variables constraining the operation of the hospital system. In addition to human inputs, a hospital must receive material inputs of money, supplies, and facilities. The success of a state hospital's interaction with the sources of political, legal, and financial support largely determine the limits of its ultimate output.

OUTPUTS

Just as the input segment of the hospital's operation is affected by its level and type of interaction with the community, so will the hospital's output be similarly influenced.

The discharge of a patient from the hospital involves such community elements as his family, employer, annd friends. Their receptivity or resistance will profoundly affect the hospital's output performance.

The major output of the present mental hospital is treated patients who can be restored to the community. In the community mental health center, however, the output will be more varied and theoretically also should include such products as changes in the level of health of the population that should result from primary and secondary prevention activities[66].

Although much attention is being devoted to the expanded community-oriented outputs of the mental hospital, unfortunately there will be a continued need to provide custody for some element of the population in the foreseeable future. As long as some patients are defined as dangerous to themselves and others, there will be a need for the protection of the patients, staff, and larger community. Because of its physical facilities, the mental hospital will be called upon to maintain some patients within its walls. This means it also will have a patient maintenance output, including all those aspects of care necessary for maintaining chronic patients within a restricted environment.

The related operating system for, on the one hand, producing rehabilitation and, on the other hand, providing maintenance and protection will come into conflict since these outcomes are in some ways incompatible. Levinson and his associates have well documented the conflict between the custodial and therapeutic orientations in the functioning of mental hospitals[67].

As an educational research institution, the hospital will produce output of trained or educated personnel and also education for the community to fulfill its goals of changing the attitudes of the community.

Another kind of output of any organization is the morale of its employees. If the hospital is unsuccessful in maintaining a reasonably high state of morale, this will feed back as a loss in the input of necessary motivation power. Like any other organization, the mental hospital has an investment in maintaining the mental and physical health of its employees.

SUBSYSTEMS OF THE ORGANIZATION

Analysis of the input and output activities of the open-system hospital in interaction with its community must be supplemented by an analysis of the interaction of the hospital's own internal system within its boundaries. These interactions within the system are between role incumbents and between subsystems.

A system can be examined in terms of its units of process and its units of structure. The totality of all the structures in a system which carry out a particular process is a **subsystem**. A subsystem thus exists in one or more distinguishable units of the total system and is identified by the process it carries out. Certain subsystems are particularly relevant for the study of the changing mental hospital.

Operating Subsystems

The subsystems which carry out the input-conversion-output processes are the "operating systems"[68]. Three types of operating systems corresponding to the three aspects of the flow process through a system can be identified in the hospital system:

1. Input operating systems, which function to take in patients, staff, money, and materials, etc.

2. Conversion operating systems, which function in treating patients in and out of the hospital, educating personnel, producing research knowledge, etc.

3. Output operating systems, which function to place treated patients in the community, place trained personnel, and disseminate knowledge, etc.

In a changing mental hospital new operating structures, such as geographically regionalized treatment units, consultation service programs, and home treatment services, must be examined in terms of their structure and functioning. By recognizing the principle of "equifinality"[69], i.e., that identical final states may be reached from different initial starting positions and by different routes, the examination of operating systems can make clear which is the more efficient of two systems set up to produce the same outcome. This is particularly important in health programs, which use a shotgun approach for providing a number of services aimed at the same targets.

Managerial Subsystem

A particularly important subsystem in the changing mental hospital is the managerial one. As the controlling or decision-making part of the organization, it cuts across all of the operating structures of production, maintenance, and adaptation. The managerial functions in the traditional hospital have been the exclusive province of the superintendent, but as the organizational structure is elaborated, the functions of management become more complex and there is pressure to share the functions of management.

In the traditional public mental hospital, there are few specialized boundary roles for linking the institution with its community. In many mental hospitals the superintendent acts as the principal link. In fact, it still is a common practice for many state hospitals to have a statement such as "address all correspondence and moneys to superintendent" printed on all stationery bearing the hospital

letterhead. As a hospital becomes a community mental health center, a need will arise for an increased number of boundary-spanning roles. The requirements of interaction with the environment will be far greater and the use of consultants to community health and welfare resources will become an increasingly common technique for bridging the gap between the hospital and its community.

In times of change it is particularly important for management to look realistically outward as well as inward and to relate the inside and outside aspects of the organization effectively. Not only must management coordinate the environmental pressures with the internal system forces, it must deal with conflicting demands of the system's substructures. Because management finds it easier to meet conflicts as they come up from day-to-day dealing with one part of the structure and then compromising with another part, the organization may be seen to jerk along in fits and starts.

In many mental hospitals the immediate crises and pressures confronting the superintendent generally emanate from within his own system and it is easy to become enmeshed by them. The ability to withstand them and focus on subtler, less immediate pressures from the external community is necessary if the hospital is to proceed in its transition from traditional functions to the expanded ones of a community mental health center. Not only does management have to function to coordinate the substructures and resolve conflicts between hierarchical levels, it must also coordinate external requirements with the system's resources and needs.

ENVIRONMENT

An open-system is defined as one into which there is a continuous flow of resources from the environment and a continuous outflow of products of the system's action back to the environment. As an open-system, an organization depends for its growth and viability upon its exchanges with the environment—that part of the physical and social world outside its boundary. The environment of the mental hospital includes the community which it serves and the other organizational systems which serve as sources of legal, political, financial, technical, and professional support.

The environment is of particular importance to a mental hospital making the transition to a community mental health center since responsibility for a specific geographical community is a basic concept in its functioning. Since there seems to be general agreement that a community mental health program requires the participation of many care-giving resources, the intersystem relations of the transitional hospital are also important.

Thompson and McEwan[70] have conceptualized organizations as ranging on a continuum of power from those that dominate their environments to those that are completely dominated. If one conceives of a community mental health suprasystem as consisting of a network of health and welfare organizations embedded in an exchange network[71], it seems that organizations must adopt strategies for coming to terms with the other organizational systems in their environment. Whether cooperative or competitive strategies are adopted has major implications for the degree of integration of services and continuity of care provided to a community.

5. TRANSACTIONS AND RESOURCE ACQUISITION

Baker states that one must consider the processes beyond a hospital-based problem to evaluate the functioning of the hospital[70]. One set of processes he suggests as a focus is the exchanges and transactions between the hospital and its environment. The following study by Sol Levine and Paul White appears in most research bibliographies dealing with organizations' processes of exchange and transactions. It is especially appropriate for this anthology, not only because of its significance in organization research, but also because it deals with the relationship of nonprofit health organizations to one another. The first several pages of the article are omitted because they reiterate concepts already covered in this unit.

EXCHANGE AS A CONCEPTUAL FRAMEWORK FOR THE STUDY OF INTERORGANIZATIONAL RELATIONSHIPS

By Sol Levine and Paul E. White

Reprinted with permission from *Administrative Science Quarterly*, Volume 5, Number 4, March 1960, pp. 586–601.

The complex of community health organizations may be seen as a system with individual organizations or system parts varying in the kinds and frequency of their relationships with one another. This system is enmeshed in ever larger systems—the community, the state, and so on.

Prevention and cure of disease constitute the ideal orientation of the health agency system, and individual agencies derive their respective goals or objectives from this larger orientation. In order to achieve its specific objectives, however, an agency must possess or control certain elements. It must have clients to serve; it must have resources in the form of equipment, specialized knowledge, or the funds with which to procure them; and it must have the services of people who can direct these resources to the clients. Few, if any, organizations have enough access to all these elements to enable them to attain their objectives fully. Under realistic conditions of element scarcity, organizations must select, on the basis of expediency or efficiency, particular functions that permit them to achieve their ends as fully as possible. By function is meant a set of interrelated services or activities that are instrumental, or believed to be instrumental,

for the realization of an organization's objectives.

Although, because of scarcity, an organization limits itself to particular functions, it can seldom carry them out without establishing relationships with other organizations of the health system. The reasons for this are clear. To fulfill its functions without relating to other parts of the health system, an organization must be able to procure the necessary elements—cases, labor services, and other resources—directly from the community or outside it. Certain classes of hospitals treating a specific disease and serving an area larger than the local community probably most nearly approximate this condition. But even in this case other organizations within the system usually control some elements that are necessary or, at least, helpful to the carrying out of its functions. These may be money, equipment, or special personnel, which are conditionally lent or given. Usually agencies are unable to obtain all the elements they need from the community or through their individual efforts and, accordingly, have to turn to other agencies to obtain additional elements. The need for a sufficient number of clients, for example, is often more efficiently met through exchanges with other organizations than through independent case-finding procedures.

Theoretically, then, were all the essential elements in infinite supply there would be little need for organizational interaction and for subscription to cooperation as an ideal. Under actual conditions of scarcity, however, interorganizational exchanges are essential to goal attainment. In sum, organizational goals or objectives are derived from general health values. These goals or objectives may be viewed as defining the organization's ideal need for elements—consumers, labor services, and other resources. The scarcity of elements, however, impels the organization to restrict its activity to limited specific functions. The fulfillment of these limited functions, in turn, requires access to certain kinds of elements, which an organization seeks to obtain by entering into exchanges with other organizations.

Interaction among organizations can be viewed within the framework of an exchange model like that suggested by Homans[71]. However, the few available definitions of exchange are somewhat limited for our purposes because they tend to be bound by economics and because their referents are mainly individual or psychological phenomena and are not intended to encompass interaction between organizational entities or larger systems[72].

We suggest the following definition of organizational exchange: **Organizational exchange is any voluntary activity between two organizations which has consequences, actual or anticipated, for the realization of their respective goals or objectives.** This definition has several advantages. First, it refers to activity in general and not exclusively to reciprocal activity. The action may be unidirectional and yet involve exchange. If an organization refers a patient to another organization which then treats him,

an exchange has taken place if the respective objectives of the two organizations are furthered by the action. Pivoting the definition on goals or objectives provides for an obvious but crucial component of what constitutes an organization. The coordination of activities of a number of individuals toward some objective or goal has been designated as a distinguishing feature of organizations by students in the field[73]. Parsons, for example, has defined an organization as a "special type of social system organized about the primacy of interest in the attainment of a particular type of system goal"[74]. That its goals or objectives may be transformed by a variety of factors and that, under some circumstances, mere survival may become primary does not deny that goals or objectives are universal characteristics of organizations.

Second, the definition widens the concept of exchange beyond the transfer of material goods and beyond gratifications in the immediate present. This broad definition of exchange permits us to consider a number of dimensions of organizational interaction that would otherwise be overlooked.

Finally, while the organizations may not be bargaining or interacting on equal terms and may even employ sanctions or pressures (by granting or withholding these elements), it is important to exclude from our definition, relationships involving physical coercion or domination; hence emphasis is on the word "voluntary" in our definition.

The elements that are exchanged by health organizations fall into three main categories: 1) referrals of cases, clients, or patients; 2) the giving or receiving of labor services, including the services of volunteer, clerical, and professional personnel, and 3) the sending or receiving of resources other than labor services, including funds, equipment, and information on cases and technical matters. Organizations have varying needs of these elements depending on their particular functions. Referrals, for example, may be seen as the delivery of the consumers of services to organizations, labor services as the human means by which the resources of the organization are made available to the consumers, and resources other than labor services as the necessary capital goods.

THE DETERMINANTS OF EXCHANGE

The interdependence of the parts of the exchange system is contingent upon three related factors: 1) the accessibility of each organization to necessary elements from sources outside the health system, 2) the objectives of the organization and particular functions to which it allocates the elements it controls, and 3) the degree to which domain consensus exists among the various organizations. An ideal theory of organizational exchange would describe the interrelationship and relative contribution of each of these factors. For the present, however, we will draw on some of our preliminary findings to suggest possible relationships among these factors and to indicate that each plays a part in affecting the exchange of elements among organizations.

Gouldner has emphasized the need to differentiate the various parts of a system in terms of their relative dependence upon other parts of the system[75]. In our terms, certain system parts are relatively dependent, not having access to elements outside the system, whereas others, which have access to such elements, possess a high degree of independence or functional autonomy. The voluntary organizations of our study (excluding hospitals) can be classified into what Sills calls either corporate or federated organizations[76]. Corporate organizations are those which delegate authority downward from the national or state level to the local level. They contrast with organizations of the federated type which delegate authority upwards—from the local to the state or national level.

It appears that local member units of corporate organizations, because they are less dependent on the local health system and can obtain the necessary elements from the community or their parent organizations, interact less with other local agencies than federated organizations. This is supported by preliminary data presented in Exhibit 1-2. It is also suggested that by carrying out their activities without entering actively into exchange relationships with other organizations, corporate organizations apparently are able to maintain their essential structure and avoid consequences resulting in the displacement of state or national goals. It may be that corporate organizations deliberately choose functions that require minimal involvement with other organizations. An examination of the four corporate organizations in our preliminary study reveals that three of them give resources to other agencies to carry out their activities, and the fourth conducts broad educational programs. Such functions are less likely to involve relationships with other organizations than the more direct service organizations, those that render services to individual recipients.

An organization's relative independence from the rest of the local health agency system and greater dependence upon a system outside the community may, at times, produce specific types of disagreements with the other agencies within the local system. This is dramatically demonstrated in the criticisms expressed toward a local community branch of an official state rehabilitation organization. The state organization, to justify its existence, has to present a successful experience to the legislators—that a minimum number of persons have been suc-

Exhibit 1-2. Weighted Rankings* of Organizations, Classified by Dependence on Four Interaction Indices

Interaction Index	Sent by	N	Voluntary		Hospitals		Official	Total interaction sent
			Corporate	Federated	Without clinics	With clinics		
Referrals	Vol. corporate	4	4.5	5	3.7	4.5	5	5
	Vol. federated	10	3	4	3.7	3	4	3
	Hosps. w/o clinics	2	4.5	3	3.7	4.5	3	4
	Hosps. w. clinics	3	1	1	1.5	2	1	1
	Official	3	2	2	1.5	1	2	2
Resources	Vol. corporate	4	5	2	1	4	5	3.5
	Vol. federated	10	4	3	3	4	4	3.5
	Hosps. w/o clinics	2	2	4.5	4.5	5	3	5
	Hosps. w. clinics	3	1	1	2	1	2	1
	Official	3	3	4.5	4.5	2	1	2
Written and verbal communication	Vol. corporate	4	5	3	2	4	5	4
	Vol. federated	10	3	1	3	3	3	2.5
	Hosps. w/o clinics	2	2	5	4.5	5	4	5
	Hosps. w. clinics	3	4	4	4.5	1	1.5	2.5
	Official	3	1	2	1	2	1.5	1
Joint activities	Vol. corporate	4	4.5	4	3	5	3.5	5
	Vol. federated	10	3	3	5	3	1	3
	Hosps. w/o clinics	2	2	5	1	2	3.5	4
	Hosps. w. clinics	3	4.5	2	2	1	5	1.5
	Official	3	1	1	4	4	2	1.5

*Note: 1 indicates highest interaction; 5 indicates lowest interaction.

cessfully rehabilitated. This means that by virture of the services the organization has offered, a certain percentage of its debilitated clients are again returned to self-supporting roles. The rehabilitative goal of the organization cannot be fulfilled unless it is selective in the persons it accepts as clients. Other community agencies dealing with seriously debilitated clients are unable to get the state to accept their clients for rehabilitation. In the eyes of these frustrated agencies the state organization is remiss in fulfilling its public goal. The state agency, on the other hand, cannot commit its limited personnel and resources to the time-consuming task of trying to rehabilitate what seem to be very poor risks. The state agency wants to be accepted and approved by the local community and its health agencies, but the state legislature and the governor, being the primary source of the agency's resources, constitute its significant reference group. Hence, given the existing definition of organizational goals and the state agency's relative independence of the local health system, its interaction with other community agencies is relatively low.

The marked difference in the interaction rank position of hospitals with outpatient clinics and those without suggests other differences between the two classes of hospitals. It may be that the two types of hospitals have different goals and that hospitals with clinics have a greater "community" orientation and are more committed to the concept of "comprehensive" care than are hospitals without clinics. However, whether or not the goals of the two types of hospitals do indeed differ, those with outpatient departments deal with population groups similar to those serviced by other agencies of the health system, that is, patients who are largely ambulatory and indigent; thus they serve patients whom other organizations may also be seeking to

serve. Moreover, hospitals with outpatient clinics have greater control over their clinic patients than over those inpatients who are the charges of private physicians, and are thereby freer to refer patients to other agencies.

The functions of an organization not only represent the means by which it allocates its elements but, in accordance with our exchange formulation, also determine the degree of dependence on other organizations for specific kinds of elements, as well as its capacity to make certain kinds of elements available to other organizations. The exchange model leads us to explain the flow of elements between organizations largely in terms of the respective functions performed by the participating agencies. Indeed, it is doubtful whether any analysis of exchange of elements among organizations which ignores differences in organizational needs would have much theoretical or practical value.

In analyzing the data from our pilot community we classified agencies on the basis of their primary health functions: resource, education, prevention, treatment, or rehabilitation. Resource organizations attempt to achieve their objectives by providing other agencies with the means to carry out their functions. The four other agency types may be conceived as representing respective steps in the control of disease. We have suggested that the primary function determines an organization's need for exchange elements. Our preliminary data reveal, as expected, that treatment organizations rate highest on number of referrals and amount of resources received and that educational organizations, whose efforts are directed toward the general public, rate low on the number of referrals (see Exhibit 1-3). This finding holds even when the larger organizations—official agencies and hospitals—are excluded and the analysis is based on the remaining voluntary

Exhibit 1-3. Weighted Rankings* of Organizations, Classified by *Function* on Four Interaction Indices

Interaction index	Received by	N	Received from					Total interaction received
			Education	Resource	Prevention	Treatment	Rehabilitation	
	Education	3	4.5	5	5	5	5	5
	Resource	5	3	4	2	4	1	3
Referrals	Prevention	5	2	1	3	2	2.5	2
	Treatment	7	1	2	1	1	2.5	1
	Rehabilitation	2	4.5	3	4	3	4	4
	Education	3	4.5	5	4	5	4.5	5
	Resource	5	1.5	3	3	4	3	3.5
Resources	Prevention	5	1.5	4	2	3	4.5	3.5
	Treatment	7	3	2	1	2	2	1
	Rehabilitation	2	4.5	1	5	1	1	2
	Education	3	4	5	4.5	5	5	5
Written and	Resource	5	3	2	2	3	2	2.5
verbal	Prevention	5	2	4	3	4	4	3
communication	Treatment	7	1	1	1	2	3	1
	Rehabilitation	2	5	3	4.5	1	1	2.5
	Education	3	4	4	1	3	4.5	4
	Resource	5	2	1	3	4	1	3
Joint	Prevention	5	1	2	2	2	3	1
activities	Treatment	7	3	3	4	1	2	2
	Rehabilitation	2	5	5	5	5	4.5	5

*Note: 1 indicates highest interaction; 5 indicates lowest interaction.

agencies of our sample. As a case in point, let us consider a health organization whose function is to educate the public about a specific disease but which renders no direct service to individual clients. If it carries on an active educational program, it is possible that some people may come to it directly to obtain information and, mistakenly, in the hope of receiving treatment. If this occurs, the organization will temporarily be in possession of potential clients whom it may route or refer to other more appropriate agencies. That such referrals will be frequent is unlikely, however. It is even less likely that the organization will receive many referrals from other organizations. If an organization renders a direct service to a client, however, such as giving X-ray examinations or polio immunizations, there is greater likelihood that it will send or receive referrals.

An organization is less limited in its function in such interagency activities as discussing general community health problems, attending agency council meetings or cooperating on some aspect of fund raising. Also, with sufficient initiative even a small educational agency can maintain communication with a large treatment organization (for example, a general hospital) through exchanges of periodic reports and telephone calls to obtain various types of information. But precisely because it is an educational agency offering services to the general public and not to individuals, it will be limited in its capacity to maintain other kinds of interaction with the treatment organization. It probably will not be able to lend or give space or equipment, and it is even doubtful that it can offer the kind of instruction that the treatment organization would seek for its staff. That the organization's function establishes the range of possibilities for exchange and that other variables exert influence within the framework established by

function is suggested by some other early findings presented in Exhibit 1-4. Organizations were classfied as direct or indirect on the basis of whether or not they provided a direct service to the public. They were also classified according to their relative prestige as rated by influential leaders in the community. Organizations high in prestige lead in the number of joint activities, and prestige seems to exert some influence on the amount of verbal and written communication. Yet it is agencies offering direct services—regardless of prestige—which lead in the number of referrals and resources received. In other words, prestige, leadership, and other organizational variables seem to affect interaction patterns within limits established by the function variable.

An obvious question is whether organizations with shared or common boards interact more with one another than do agencies with separate boards. Our preliminary data show that the interaction rate is not affected by shared board membership. We have not been able to ascertain if there is any variation in organizational interaction when the shared board positions are occupied by persons with high status or influence. In our pilot community, there was only one instance in which two organizations had the same top community leaders as board members. If boards play an active role in the activities of health organizations, they serve more to link the organization to the community and the elements it possesses than to link the organization to other health and welfare agencies. The board probably also exerts influence on internal organizational operations and on establishing or approving the primary objective of the organization. Once the objective and the implementing functions are established, these functions tend to exert their influence autonomously on organizational interaction.

Exhibit 1-4. Weighted Rankings* of Organizations, Classified by *Prestige of Organization* and by *General Type of Service Offered* on Four Interaction Indices

Interaction	Received by	N	Received from					Total received
			High Prestige		Low Prestige			
Referrals	High direct	9	1	1	1	1		1
	High indirect	3	3	3.5	3	3.5		3
	Low direct	6	2	2	2	2		2
	Low indirect	4	4	3.5	4	3.5		4
Resources	High direct	9	2	2	2	2		2
	High indirect	3	3	3	3	3.5		3
	Low direct	6	1	1	1	1		1
	Low indirect	4	4	4	4	3.5		4
Written and verbal communication	High direct	9	2	2	3	1		2
	High indirect	3	3	3	1	3		3
	Low direct	6	1	1	2	2		1
	Low indirect	4	4	4	4	4		4
Joint activities	High direct	9	1	1.5	2	2		2
	High indirect	3	2	1.5	1	1		1
	Low direct	6	4	3	3	4		3
	Low indirect	4	3	4	4	3		4

*Note: 1 indicates highest interaction; 5 indicates lowest interaction.

ORGANIZATIONAL DOMAIN

As we have seen, the elements exchanged are cases, labor services, and other resources. All organizational relationships directly or indirectly involve the flow and control of these elements. Within the local health agency system, the flow of elements is not centrally coordinated but rests upon voluntary agreements or understanding. Obviously, there will be no exchange of elements between two organizations that do not know of each other's existence or that are completely unaware of each other's functions. Even more, there can be no exchange of elements without some agreement or understanding, however implicit. These exchange agreements are contingent upon the organization's domain. The domain of an organization consists of the specific goals it wishes to pursue and the functions it undertakes in order to implement its goals. In operational terms, organizational domain in the health field refers to the claims that an organization stakes out for itself in terms of 1) disease covered, 2) population served, and 3) services rendered. The goals of the organization constitute in effect the organization's claim to future functions and to the elements requisite to these functions, whereas the present or actual functions carried out by the organization constitute *de facto* claims to these elements. Exchange agreements rest upon prior consensus regarding domain. Within the health agency system, consensus regarding an organization's domain must exist to the extent that parts of the system will provide each agency with the elements necessary to attain its ends.

Once an organization's goals are accepted, domain consensus continues as long as the organization fulfills the functions adjudged appropriate to its goals and adheres to certain standards of quality. Our data show that organizations find it more difficult to legitimate themselves before other organizations in the health system than before such outside systems as the community or state. An organization can sometimes obtain sufficient elements from outside the local health system, usually in the form of funds, to continue in operation long after other organizations within the system have challenged its domain. Conversely, if the goals of a specific organization are accepted within the local agency system, other organizations of the system may encourage it to expand its functions and to realize its goals more fully by offering it elements to implement them. Should an organization not respond to this encouragement, it may be forced to forfeit its claim to the unrealized aspect of its domain.

Within the system, delineation of organizational domains is highly desired[77]. For example, intense competition may occur occasionally between two agencies offering the same services, especially when other agencies have no specific criteria for referring patients to one rather than the other. If both services are operating near capacity, competition between the two tends to be less keen, the choice being governed by the availability of service. If the services are being operated at less than capacity, competition and conflict often occur. Personnel of referring agencies in this case frequently deplore the "duplication of services" in the community. In most cases the conflict situation is eventually resolved by agreement on the part of the competing agencies to specify the criteria for referring patients to them. The agreement may take the form of consecutive handling of the same patients. For example, age may be employed as a criterion. In one case three agencies were involved in giving rehabilitation services: one took preschool children, another school children, and the third adults. In another case, where preventive services were offered, one agency took preschool children and the other took children of school age. The relative accessibility of the agencies to the respective age groups was a partial basis for these divisions. Another criterion—disease stage—also permits consecutive treatment of patients. One agency provided physical therapy to beridden patients; another handled them when they became ambulatory.

Several other considerations, such as priorities in allocation of elements, may impel an organization to delimit its functions even when no duplication of services exists. The phenomenon of delimiting one's role and consequently of restricting one's domain is well known. It can be seen, for instance, in the resistance of certain universities of high prestige to offer "practical" or vocational courses, or courses to meet the needs of any but high-status professionals, even to the extent of foregoing readily accessible federal grants. It is evidenced in the insistence of certain psychiatric clinics on handling only cases suitable for psychoanalytic treatment, of certain business organizations on selling only to wholesalers, of some retail stores on handling only expensive merchandise.

The flow of elements in the health system is contingent upon solving the problem of "who gets what for what purpose." The clarification of organizational domains and the development of greater domain consensus contributes to the solution of this problem. In short, domain consensus is a prerequisite to exchange. Achieving domain consensus may involve negotiation, orientation, or legitimation. When the functions of the interacting organizations are diffuse, achieving domain consensus becomes a matter of constant readjustment and compromise, a process which may be called negotiation or bargaining. The more specific the functions, however, the more domain consensus is attained merely by orientation (for example, an agency may call an X-ray unit to inquire about the specific procedures

for implementing services). A third, less frequent but more formalized, means of attaining domain consensus is the empowering, licensing, or "legitimating" of an organization to operate within the community by some other organization. Negotiation, as a means of attaining domain consensus seems to be related to diffuseness of function, whereas orientation, at the opposite extreme, relates to specificity of function.

These processes of achieving domain consensus constitute much of the interaction between organizations. While they may not involve the immediate flow of elements, they are often necessary preconditions for the exchange of elements, because without at least minimal domain consensus there can be no exchange among organizations. Moreover, to the extent that these processes involve proferring information about the availability of elements as well as about rights and obligations regarding the elements, they constitute a form of interorganizational exchange.

DIMENSIONS OF EXCHANGE

We have stated that all relationships among local health agencies may be conceptualized as involving exchange. There are four main dimensions to the actual exchange situation. They are:

1. **The parties to the exchange.** The characteristics we have thus far employed in classifying organizations or the parties to the exchange are organizational form or affiliation, function, prestige, size, personnel characteristics, and numbers and types of clients served.

2. **The kinds and quantities exchanged.** These involve two main classes: the actual elements exchanged (consumers, labor services, and resources other than labor services) and information on the availability of

these organizational elements and on rights and obligations regarding them.

3. The agreement underlying the exchange. Every exchange is contingent upon a prior agreement, which may be implicit and informal or fairly explicit and highly formalized. For example, a person may be informally routed or referred to another agency with the implicit awareness or expectation that the other organization will handle the case. On the other hand, the two agencies may enter into arrangements that stipulate the exact conditions and procedures by which patients are referred from one to another. Furthermore, both parties may be actively involved in arriving at the terms of the agreement, or these terms may be explicitly defined by one for all who may wish to conform to them. An example of the latter case is the decision of a single organization to establish a policy of a standard fee for service.

4. The direction of the exchange. This refers to the direction of the flow of organizational elements. We have differentiated three types:

a. **Unilateral,** where elements flow from one organization to another and no elements are given in return.

b. **Reciprocal,** where elements flow from one organization to another in return for other elements.

c. **Joint,** where elements flow from two organizations acting in unison toward a third party. This type, although representing a high order of agreement and coordination of policy among agencies, does not involve the actual transfer of elements.

As we proceed with our study of relationships among health agencies, we will undoubtedly modify and expand our theoretical model. For example, we will attempt to describe how the larger systems are intertwined with the health agency system. Also, we will give more attention to the effect of interagency competition and conflict regarding the flow of elements among organizations. In this respect we will analyze differences among organizations with respect not only to domain but to fundamental goals as well. As part of this analysis we will examine the orientations of different categories of professionals (for example, nurses and social workers) as well as groups with varying experiences and training within categories of professionals (as nurses with or without graduate education).

In the meantime, we find the exchange framework useful in ordering our data, locating new areas for investigation, and developing designs for studying interorganizational relationships. We feel that the conceptual framework and findings of our study will be helpful in understanding not only health agency interaction but also relationships within other specific systems (such as military, industrial, governmental, educational, and other systems). As in our study of health agencies, organizations within any system may confidently be expected to have need for clients, labor, and other resources. We would also expect that the interaction pattern among organizations within each system will also be affected by 1) organizational function, 2) access to the necessary elements from outside the system, and 3) the degree of domain consensus existing among the organizations of the system. It appears that the framework also shows promise in explaining interaction among organizations belonging to different systems (for example, educational and business systems, educational and governmental, military and industrial, and so forth). Finally, we believe our framework has obvious value in explaining interaction among units or departments within a single large-scale organization.

Levine and White described how nonprofit health organizations vie for resources and referrals. Perhaps one of the greatest contributions of modern organization theory is its insistence that the manager look at the nature of an organization's relationship with its environment for a true measure of its effectiveness. In line with this assertion, Yuchtman and Seashore[78] suggest that complex organizations exist for the benefit of society; thus, society is the appropriate frame of reference for the evaluation of organizational effectiveness. However, taking the organization itself as the frame of reference, its contribution to the larger system must be regarded as an unavoidable and costly requirement rather than a sign of success.

Yuchtman and Seashore define organization effectiveness in terms of the bargaining position of the organization the extent to which the organization has the ability to enter into competition with other organizations and acquire the scarce and valued resources for which they compete.

Resources are defined by Yuchtman as "generalized means or facilities that are potentially controllable by social organizations and are potentially usable in the relationships between the organization and its environment." Gamson[79] states that the reputation of groups influential in community affairs is a resource. Presthus[80] identifies political resources which fit well into Yuchtman's definition: expertise, experience, power, information, constituency support, prestige. Other resources for which institutions compete include favorable legislation, referrals, funding, contracts, clientele, employees, technology, and community reputation.

The better the bargaining position of an organization, the more capable it is of attaining its organizational goals and the more capable it is of allowing its members to attain their personal goals. To use a simple example, if a given hospital has difficulty competing with other hospitals in the recruitment of professional nursing staff, it can be assumed that the hospital has a poor bargaining position in relation to the other users in the community vying for the same scarce resource. The quality of care may suffer because there are not enough nurses to cover the patient load, or the hospital may be forced to close beds, thereby decreasing revenues. Chronic shortage of nursing personnel not only compromises the institutional goals of high-quality patient care and fiscal solvency, but the personal goals of the nursing personnel suffer. Overtaxed nurses are unable to have the satisfaction of work completed and well done; they have difficulty mustering pride in their institution when they must continually apologize for undelivered services; their professional growth is stunted through lack of time for patient conferences, inservice training, and clinical discussions with other professionals. Word gets out in the community about the difficult working conditions as compared to other institutions which have good staffing. The hospital with shortages then may have even more difficulty recruiting new personnel; the bargaining position may worsen. A number of factors may interfere with a hospital's ability to bargain for nurses in an environment where the supply is scarce. The hospital may be located in a deteriorating community where personal safety is a concern, medical staff may have a derogatory attitude about nurses' clinical competencies, and administration may be unwilling to pay competitive salaries. If the institution undertakes actions to correct the problems it can be said to be adopting strategies or formulating organizational goals which will enhance its bargaining position. What determines the success of its actions still remains outside the institution, for success ultimately will be measured in terms of its ability to compete, its ability to improve its bargaining position in the community. Thus, an organization's goal is not the ultimate criterion of organization effectiveness; rather, effectiveness is considered in the more general capability of the organization as a resource-getting system. As Yuchtman and Seashore state:

By focusing on the ability of the organization to exploit its environment in the acquisition of resources, we are directed by the basic yet often neglected fact that it is only in the arena of competition over scarce and valued resources that the performance of both like and unlike organizations can be assessed and evaluated comparatively. To put it somewhat differently, any change in the relation between the organization and its environment is affected by and results in a better or worse bargaining position vis-a-vis that environment or parts thereof.[81]

For Yuchtman and Seashore, exchange and competition are at the extreme ends of a continuum along which input and output transactions of the institution can be described. Change implies cooperation and agreement; competition implies rivalry and capitalizing on social, political, or economic advantage.

Organizations commit considerable energy to activities which will enhance their power to acquire certain resources, or they shift among a class of resources according to their degree of relevancy. For example, a hospital will acquire the latest in medical technology in order to compete with other hospitals for physicians who will fill the hospital's beds, or a community mental health center will develop inpatient beds and medical care status in order to compete with general hospitals for psychiatric patients with medical problems.

Organizations compete with one another for more overt types of political control. Health care providers in institutions frequently seek representation on state rate review boards or health systems agency committees. Hospital trustees actively recruit new members who have community influence. I am sure you can identify other strategies health professionals use to enhance or to maintain the bargaining position of the organizations to which they belong.

GENERAL SYSTEMS THEORY: APPLICATIONS FOR ORGANIZATION AND MANAGEMENT

By Fremont E. Kast and James E. Rosenweig

Reprinted with permission from *Academy of Management Journal*, December 1972, pp. 447–464. Published by Academy of Management Publications.

Thompson[82] states that the essence of administration lies in understanding basic configurations that exist between subsystems (a set of like organizations) and their external environment. The selections which have been presented thus far provide the nurse administrator with the background and open systems constructs upon which most of the literature on organization-environment relationships is based. Nurse administrators who familiarize themselves with these conceptual tools should be able to identify the rationale for many of the actions their institutions undertake as they interact with other organizational systems of the larger environment.

Kast and Rosenzweig's treatment of applications of general systems theory rounds out this discussion of the open-systems paradigm and modern organization theory. The selection is an excellent summary of open-systems conceptualizations and their limitations. The authors mention contingency views as a step towards less abstraction. The contingency approach is discussed more extensively in the next unit as the effects of task environments on organization structure and effectiveness are considered.

Biological and social scientists generally have embraced systems concepts. Many organization and management theorists seem anxious to identify with this movement and to contribute to the development of an approach which purports to offer the ultimate—the unification of all science into one grand conceptual model. Who could possibly resist? General systems theory seems to provide a relief from the limitations of more mechanistic approaches and a rationale for rejecting "principles" based on relatively "closed-system" thinking. This theory provides the paradigm for organization and management theorists to "crank into their systems model" all of the diverse knowledge from relevant underlying disciplines. It has become almost mandatory to have the word "system" in the title of recent articles and books (many of us have compromised and placed it only in the subtitle)[83].

But where did it all start? This question takes us back into history and brings to mind the long-standing philosophical arguments between mechanistic and organismic models of the nineteenth and early twentieth centuries. As Deutsch says:

Both mechanistic and organismic models were based substantially on experiences and operations known before 1850. Since then, the experience of almost a century of scientific and technological progress has so far not been utilized for any significant new model for the study of organization and in particular of human thought.[84]

General systems theory even revives the specter of the "vitalists" and their views on "life force" and most certainly brings forth renewed questions of teleological or purposeful behavior of both living and nonliving systems. Phillips and others have suggested that the philosophical roots of general systems theory go back even further, at least to the German philosopher Hegel (1770–1831)[85]. Thus, we should recognize that in the adoption of the systems approach for the study of organizations we are not dealing with newly discovered ideas—they have a rich geneaology.

Even in the field of organization and management theory, systems views are not new. Chester Barnard used a basic systems framework:

A cooperative system is a complex of physical, biological, personal, and social components which are in a specific systematic relationship by reason of the cooperation of two or more persons for at least one definite end. Such a system is evidently a subordinate unit of larger systems from one point of view; and itself embraces subsidiary systems—physical, biological, etc.—from another point of view. One of the systems comprised within a cooperative system, the one which is implicit in the phrase "cooperation of two or more persons," is called an "organization."[86]

And Barnard was influenced by the "systems views" of Vilfredo Pareto and Talcott Parsons. Certainly this quote (dressed up a bit to give the term "system" more emphasis) could be the introduction to a 1972 book on organizations.

Miller points out that Alexander Bogdanov, the Russian philosopher, developed a theory of "tektology" or universal organization science in 1912 which foreshadowed general systems theory and used many of the same concepts as modern systems theorists[87].

However, in spite of a long history of organismic and holistic thinking, the utilization of the systems approach did not become the accepted model for organization and management writers until relatively recently. It is difficult to specify the turning point exactly. The momentum of systems thinking was identified by Scott in 1961 when he described the relationship between general systems theory and organization theory.

The distinctive qualities of modern organization theory are its conceptual-analytical base, its reliance on empirical research data, and above all, its integrating nature. These qualities are framed in a philosophy which accepts the premise that the only meaningful way to study organization is to study it as a system . . . Modern organization theory and general system theory are similar in that they look at organization as an integrated whole.[88]

Scott said explicitly what many in our field had been thinking and/or implying—he helped us put into perspective the important writings of Herbert Simon, James March, Talcott Parsons, George Homans, E. Wight Bakke, Kenneth Boulding, and many others.

But how far have we really advanced over the past decade in applying general systems theory to organizations and their management? Is it still a "skeleton" or have we been able to "put some meat on the bones"? The systems approach has been touted because of its potential usefulness in understanding the complexities of "live" organizations. Has this approach really helped us in this endeavor or has it compounded confusion with chaos? Herbert Simon describes the challenge for the systems approach:

In both science and engineering, the study of "systems" is an increasingly popular activity. Its popularity is more a response to a pressing need for synthesizing and analyzing complexity than it is to any large development of a body of knowledge and technique for dealing with complexity. If this popularity is to be more than a fad, necessity will have to mother invention and provide substance to go with the name.[89]

In this article we will explore the issue of whether we are providing substance for the term *systems approach* as it relates to the study of organizations and their management. There are many interesting historical and philosophical questions concerning the relationship between the mechanistic and organistic approaches and their applicability to the various fields of science, as well as other interesting digressions into the evolution of systems approaches. However, we will resist those temptations and plunge directly into a discussion of the key concepts of general systems theory, the way in which these ideas have been used by organization theorists, the limitations in their application, and some suggestions for the future.

KEY CONCEPTS OF
GENERAL SYSTEMS THEORY

The key concepts of general systems theory have been set forth by many writers and have been used by many organization and management theorists. It is not our purpose here to elaborate on them in great detail because we anticipate that most readers will have been exposed to them in some depth. Exhibit 1-5 provides a very brief review of those characteristics of systems which seem to have wide acceptance. The review is far from complete. It is difficult to identify a complete list of characteristics derived from general systems theory; moreover, it is merely a first-order classification. There are many derived second- and third-order characteristics which could be considered.

For example, James G. Miller sets forth 165 hypotheses, stemming from open systems theory, which might be applicable to two or more levels of systems[92]. He suggests that they are **general** systems theoretical hypotheses and qualifies them by suggesting that they are propositions applicable to general systems **behavior** theory and would thus exclude nonliving systems. He does not limit these propositions to individual organisms, but considers them appropriate for social systems as well. His hypotheses are related to such issues as structure, process, subsystems, information, growth and integration. It is obviously impossible to discuss all of these hypotheses; we want only to indicate the extent to which many interesting propositions are being posed which might have relevance to many different types of systems. It will be a very long time (if ever) before most of these hypotheses are validated; however, we are surprised at how many of them can be agreed with intuitively, and we can see their possible verification in studies of social organizations.

We turn now to a closer look at how successful or unsuccessful we have been in utilizing these concepts in the development of modern organization theory.

A BEGINNING: ENTHUSIASTIC
BUT INCOMPLETE

We have embraced general systems theory but, really, how completely? We could review a vast literature in modern organization theory which has explicitly or implicitly adopted systems theory as a frame of reference, and we have investigated in detail a few representative examples of the literature in assessing the state of the art[93]. It was found that most of these books professed to utilize general systems theory. Indeed, in the first few chapters, many of them did an ex-

Exhibit 1-5.　Key Concepts of General Systems Theory

Subsystems or Components. A system by definition is composed of interrelated parts or elements. This is true for all systems—mechanical, biological, and social. Every system has at least two elements, and these elements are interconnected.

Holism, Synergism, Organicism, and Gestalt. The whole is not just the sum of the parts; the system itself can be explained only as a totality. Holism is the opposite of elementarism, which views the total as the sum of its individual parts.

Open-Systems View. Systems can be considered in two ways: 1) closed or 2) open. Open systems exchange information, energy, or material with their environments. Biological and social systems are inherently open systems; mechanical systems may be open or closed. The concepts of open and closed systems are difficult to defend in the absolute. We prefer to think of open-closed as a dimension; that is, systems are relatively open or relatively closed.

Input-Transformation-Output Model. The open system can be viewed as a transformation model. In a dynamic relationship with its environment, it receives various inputs, transforms these inputs in some way, and exports outputs.

System Boundaries. It follows that systems have boundaries which separate them from their environments. The concept of boundaries helps us understand the distinction between open and closed systems. The relatively closed system has rigid, impenetrable boundaries; whereas the open system has permeable boundaries between itself and a broader suprasystem. Boundaries are relatively easily defined in physical and biological systems, but are very difficult to delineate in social systems such as organizations.

Negative Entropy. Closed physical systems are subject to the force of entropy which increases until eventually the entire system fails. The tendency toward maximum entropy is a movement to disorder, complete lack of resource transformation, and death. In a closed system, the change in entropy must always be positive; however, in open biological or social systems, entropy can be arrested and may even be transformed into negative entropy—a process of more complete organization and ability to transform resources—because the system imports resources from its environment.

Steady State, Dynamic Equilibrium, and Homeostasis. The concept of steady state is closely related to that of negative entropy. A closed system eventually must attain an equilibrium state with maximum entropy—death or disorganization. However, an open system may attain a state where the system remains in dynamic equilibrium through the continuous inflow of materials, energy, and information.

Feedback. The concept of feedback is important in understanding how a system maintains a steady state. Information concerning the outputs or the process of the system is fed back as an input into the system, perhaps leading to changes in the transformation process and/or future outputs. Feedback can be both positive and negative, although the field of cybernetics is based on negative feedback. Negative feedback is informational input which indicates that the system is deviating from a prescribed course and should readjust to a new steady state.

Hierarchy. A basic concept in systems thinking is that of hierarchical relationships between systems. A system is composed of subsystems of a lower order and is also part of a suprasystem. Thus, there is a hierarchy of the components of the system.

Internal Elaboration. Closed systems move toward entropy and disorganization. In contrast, open systems appear to move in the direction of greater differentiation, elaboration, and a higher level of organization.

Multiple Goal-Seeking. Biological and social systems appear to have multiple goals or purposes. Social organizations seek multiple goals, if for no other reason than that they are composed of individuals and subunits with different values and objectives.

Equifinality of Open Systems. In mechanistic systems there is a direct cause and effect relationship between the initial conditions and the final state. Biological and social systems operate differently. Equifinality suggests that certain results may be achieved with different initial conditions and in different ways. This view suggests that social organizations can accomplish their objectives with diverse inputs and with varying internal activities (conversion processes).

cellent job of presenting basic systems concepts and showing their relationship to organizations; however, when they moved further into the discussion of more specific subject matter, they departed substantially from systems theory. The studies appear to use a ''partial systems approach'' and leave for the reader the problem of integrating the various ideas into a systemic whole. It also appears that many of the authors are unable, because of limitations of knowledge about subsystem relationships, to carry out the task of using general systems theory as a conceptual basis for organization theory.

Furthermore, it is evident that each author had many ''good ideas'' stemming from the existing body of knowledge or current research on organizations which did not fit neatly into a ''systems model.'' For example, they might discuss leadership from a relatively closed-system point of view and not consider it in relation to organizational technology, structure, or other variables. Our review of the literature suggests that much remains to be done in applying general systems theory to organization theory and management practice.

SOME DILEMMAS IN APPLYING GST TO ORGANIZATIONS

Why have writers embracing general systems theory as a basis for studying organizations had so much difficulty in following through? Part of this difficulty may stem from the newness of the paradigm and our inability to operationalize all we think we know about this approach. Or it may be because we know too little about the systems under investigation. Both of these possibilities will be covered later, but first we need to look at some of the more specific conceptual problems.

Organizations as Organisms

One of the basic contributions of general systems theory was the rejection of the traditional closed-system or mechanistic view of social organizations. But, did general systems theory free us from this constraint only to impose another, less obvious one? General systems theory grew out of the organismic views of von Bertalanffy and other biologists; thus, many of the characteristics are relevant to the living organism. It is conceptually easy to draw the analogy between living organisms and social organizations. "There is, after all, an intuitive similarity between the organization of the human body and the kinds of organizations men create. And so, undaunted by the failures of the human-social analogy through time, new theorists try afresh in each epoch"[94]. General systems theory would have us accept this analogy between organism and social organization. Yet, we have a hard time swallowing it whole. Katz and Kahn warn us of the danger:

There has been no more pervasive, persistent, and futile fallacy handicapping the social sciences than the use of the physical model for the understanding of social structures. The biological metaphor, with its crude comparisons of the physical parts of the body to the parts of the social system, has been replaced by more subtle but equally misleading analogies between biological and social functioning. This figurative type of thinking ignores the essential difference between the socially contrived nature of social systems and the physical structure of the machine or the human organism. So long as writers are committed to a theoretical framework based upon the physical model, they will miss the essential social-psychological facts of the highly variable, loosely articulated character of social systems. [95]

In spite of this warning, Katz and Kahn do embrace much of the general systems theory concepts which are based on the biological metaphor. We must be very cautious about trying to make this analogy too literal. We agree with Silverman, who says, "It may, therefore, be necessary to drop the analogy between an organization and an organism: organizations may be systems but not necessarily **natural** systems"[96].

Distinction between Organization and an Organization

General systems theory emphasizes that systems are organized—they are composed of interdependent components in some rela-

tionship. The social organization would then follow logically as just another system. But, we are perhaps being caught in circular thinking. It is true that all systems (physical, biological, and social) are by definition organized, but are all systems organizations? Rapoport and Horvath distinguish "organization theory" and "the theory of organizations" as follows:

We see organization theory as dealing with general and abstract organizational principles: it applies to any system exhibiting organized complexity. As such, organization theory is seen as an extension of mathematical physics or, even more generally, of mathematics designed to deal with organized systems. The theory of organizations, on the other hand, purports to be a social science. It puts real human organizations at the center of interest. It may study the social structure of organizations and so can be viewed as a branch of sociology; it can study the behavior of individuals or groups as members of organizations and can be viewed as a part of social psychology; it can study power relations and principles of control in organizations and so fits into political science.[97]

Why make an issue of this distinction? It seems to us that there is a vital matter involved. All systems may be considered to be organized, and more advanced systems may display differentiation in the activities of component parts—such as the specialization of human organs. However, all systems **do not** have purposeful entities. Can the heart or lungs be considered as purposeful entities in themselves or are they only components of the larger purposeful system, the human body? By contrast, the social organization is composed of two or more purposeful elements. "An organization consists of elements that have and can exercise their own wills"[98]. Organisms, the foundation stone of general systems theory, do not contain purposeful elements which exercise their own will. This distinction between the organism and the social organization is of importance. In much of general systems theory, the concern is primarily with the way in which the **organism** responds to environmentally generated inputs. Feedback concepts and the maintenance of a steady state are based on internal adaptations to environmental forces. (This is particularly true of cybernetic models.) But, what about those changes and adaptations which occur from **within** social organizations? Purposeful elements within the social organization may initiate activities and adaptations which are difficult to subsume under feedback and steady-state concepts.

Open and Closed Systems

Another dilemma stemming from general systems theory is the tendency to dichotomize all systems as opened or closed. We have been led to think of physical systems as closed, subject to the laws of entropy, and to think of biological systems as open to their environment and, possibly, becoming negentropic. But applying this strict polarization to social organizations creates many difficulties. In fact, most social organizations and their subsystems are "partially open" and "partially closed." Open and closed are a matter of degree. Unfortunately, there seems to be a widely held view (often more implicit than explicit) that **open-system thinking is good** and **closed-system thinking is bad.** We have not become sufficiently sophisticated to recognize that both are appropriate under certain conditions. For example, one of the most useful conceptualizations set forth by Thompson is that the social organization **must seek** to use closed-system concepts (particularly at the technical core) to reduce uncertainty and to create more effective performance at this level.

Still Subsystems Thinking

Even though we preach a general systems approach, we often practice subsystems think-

ing. Each of the academic disciplines and each of us personally have limited perspective of the system we are studying. While proclaiming a broad systems viewpoint, we often dismiss variables outside our interest or competence as being irrelevant, and we only open our system to those inputs which we can handle with our disciplinary bag of tools. We are hampered because each of the academic disciplines has taken a narrow "partial systems view" and find comfort in the relative certainty which this creates. Of course, this is not a problem unique to modern organization theory. Under the more traditional process approach to the study of management, we were able to do an admirable job of delineating and discussing planning, organizing, and controlling as separate activities. We were much less successful in discussing them as integrated and interrelated activities.

How Does Our Knowledge Fit?

One of the major problems in utilizing general systems theory is that we know (or think we know) more about certain relationships than we can fit into a general systems model. For example, we are beginning to understand the two-variable relationship between technology and structure. But, when we introduce another variable, say psychosocial relationships, our models become too complex. Consequently, in order to discuss all the things we know about organizations, we depart from a systems approach. Perhaps it is because we know a great deal more about the elements or subsystems of an organization than we do about the interrelationships and interactions between these subsystems. And, general systems theory forces us to consider those relationships about which we know the least—a true dilemma. So we continue to elaborate on those aspects of the organization which we know best—a partial systems view.

Failure to Delineate a Specific System

When the social sciences embraced general systems theory, the total system became the focus of attention and terminology tended toward vagueness. In the utilization of systems theory, we should be more precise in delineating the specific system under consideration. Failure to do this leads to much confusion. As Murray suggests:

I am wary of the word "system" because social scientists use it very frequently without specifying which of several possible different denotations they have in mind; but more particularly because, today, "system" is a highly cathected term, loaded with prestige; hence, we are all strongly tempted to employ it even when we have nothing definite in mind and its only service is to indicate that we subscribe to the general premise respecting the interdependence of things—basic to organismic theory, holism, field theory, interactionism, transactionism, etc. . . . When definitions of the units of a system are lacking, the term stands for no more than an article of faith, and is misleading to boot, insofar as it suggests a condition of affairs that may not actually exist.[99]

We need to be much more precise in delineating both the boundaries of the system under consideration and the level of our analysis. There is a tendency for current writers in organization theory to accept general systems theory and then to move indiscriminately across systems boundaries and between levels of systems without being very precise (and letting their readers in on what is occurring). James Miller suggests the need for clear delineation of levels in applying systems theory: "It is important to follow one procedural rule in systems theory in order to avoid confusion. Every discussion should begin with an identification of the level of reference, and the discourse should not change to another level without a specific statement that this is occur-

ring"[100]. Our field is replete with these confusions about systems levels. For example, when we use the term "organizational behavior" are we talking about the way the organization behaves as a system or are we talking about the behavior of the individual participants? By goals, do we mean the goals of the organization or the goals of the individuals within the organization? In using systems theory we must become more precise in our delineation of systems boundaries and systems levels if we are to prevent confusing conceptual ambiguity.

Recognition That Organizations Are "Contrived Systems"

We have a vague uneasiness that general systems theory truly does not recognize the "contrived" nature of social organizations. With its predominant emphasis on natural organisms, it may understate some characteristics which are vital for the social organization. Social organizations do not occur naturally in nature; they are contrived by man. They have structure; but it is the structure of events rather than of physical components, and it cannot be separated from the processes of the system. The fact that social organizations are contrived by human beings suggests that they can be established for an infinite variety of purposes and do not follow the same life-cycle patterns of birth, growth, maturity, and death as biological systems. As Katz and Kahn say:

Social structures are essentially contrived systems. They are made of men and are imperfect systems. They can come apart at the seams overnight, but they can also outlast by centuries the biological organisms which originally created them. The cement which holds them together is essentially psychological rather than biological. Social systems are anchored in the attitudes, perceptions, beliefs, motivations, habits, and expectations of human beings.[101]

Recognizing that the social organization is contrived again cautions us against making an exact analogy between it and physical or biological systems.

Questions of Systems Effectiveness

General systems theory with its biological orientation would appear to have an evolutionary view of system effectiveness. That living system which best adapts to its environment prospers and survives. The primary measure of effectiveness is perpetuation of the organism's species. Teleological behavior is therefore directed toward survival. But, is survival the only criterion of effectiveness of the social system? It is probably an essential but not all-inclusive measure of effectiveness.

General systems theory emphasizes the organism's survival goal and does not fully relate to the question of the effectiveness of the system in its suprasystem—the environment. Parsonian functional-structural views provide a contrast. "The **raison d'etre** of complex organizations, according to this analysis, is mainly to benefit the society in which they belong, and that society is, therefore, the appropriate frame of reference for the evaluation of organizational effectiveness"[102].

But, this view seems to go to the opposite extreme from the survival view of general systems theory—the organization exists to serve the society. It seems to us that the truth lies somewhere between these two viewpoints. And it is likely that a systems viewpoint (modified from the species survival view of general systems theory) will be most appropriate. Yuchtman and Seashore suggest:

The organization's success over a period of time in this competition for resources—i.e., its bargaining position in a given environment—is regarded as an expression of its overall effectiveness. Since the resources are of various kinds, and the competitive relationships are multiple, and since there is interchangeability among classes of resources, the assessment of organizational effectiveness must be in terms not of any single criterion but of an open-ended multidimensional set of criteria.[103]

This viewpoint suggests that questions of organizational effectiveness must be concerned with at least three levels of analysis. The level of the environment, the level of the social organization as a system, and the level of the subsystems (human participants) within the organization. Perhaps much of our confusion and ambiguity concerning organizational effectiveness stems from our failure to clearly delineate the level of our analysis and, even more important, our failure really to understand the relationships among these levels.

Our discussion of some of the problems associated with the application of general systems theory to the study of social organizations might suggest that we completely reject the appropriateness of this model. On the contrary, we see the systems approach as the new paradigm for the study of organizations; but, like all new concepts in the sciences, one which has to be applied, modified, and elaborated to make it as useful as possible.

SYSTEMS THEORY PROVIDES THE NEW PARADIGM

We hope the discussion of general systems theory and organizations provides a realistic appraisal. We do not want to promote the value of the systems approach as a matter of faith; however, we do see systems theory as vital to the study of social organizations and as providing the major new paradigm for our field of study.

Thomas Kuhn provides an interesting interpretation of the nature of scientific revolution[104]. He suggests that major changes in all fields of science occur with the development of new conceptual schemes, or paradigms. These new paradigms do not just represent a step-by-step advancement in "normal" science (the science generally accepted and practiced) but, rather, a revolutionary change in the way the scientific field is perceived by the practitioners. Kuhn says:

The historian of science may be tempted to exclaim that when paradigms change, the world itself changes with them. Led by a new paradigm, scientists adopt new instruments and look in new places. Even more important, during revolutions scientists see new and different things when looking with familiar instruments in places they have looked before. It is rather as if the professional community has been suddenly transported to another planet where familiar objects are seen in a different light and are joined by unfamiliar ones as well. . . . Paradigm changes do cause scientists to see the world of their research-engagement differently. Insofar as their only recourse to that world is through what they see and do, we may want to say that after a revolution scientists are responding to a different world.[105]

New paradigms frequently are rejected by the scientific community. (At first they may seem crude and limited—offering very little more than older paradigms.) They frequently lack the apparent sophistication of the older paradigms, which they ultimately replace. They do not display the clarity and certainty of older paradigms, which have been refined through years of research and writing. But, a new paradigm does provide for a "new start" and opens up new directions which were not possible under the old. "We must recognize how very limited in both scope and precision a paradigm can be at the time of its first appearance. Paradigms gain their status because they are more successful than their competitors in solving a few problems that the group of practitioners has come to recognize as acute. To be more successful is not, however, to be either completely successful with a single problem or

notably successful with any large number"[106].

Systems theory does provide a new paradigm for the study of social organizations and their management. At this stage it is obviously crude and lacking in precision. In some ways it may not be much better than older paradigms which have been accepted and used for a long time (such as the management process approach). As in other fields of scientific endeavor, the new paradigm must be applied, clarified, elaborated, and made more precise. But, it does provide a fundamentally different view of the reality of social organizations and can serve as the basis for major advancements in our field.

We see many exciting examples of the utilization of the new systems paradigm in the field of organization and management. Several of these have been referred to earlier, and there have been many others. Burns and Stalker made substantial use of systems views in setting forth their concepts of mechanistic and organic managerial systems[108]. Their studies of the characteristics of these two organization types lack precise definition of the variables and relationships, but their colleagues have used the systems approach to look at the relationship of organizations to their environment and also among the technical, structural, and behavioral characteristics within the organization[109]. Chamberlain used a system view in studying enterprises and their environment which is substantially different from traditional microeconomics[110]. The emerging field of "environmental sciences" and "environmental administration" has found the systems paradigm vital.

Thus, the systems theory paradigm is being used extensively in the investigation of relationships between subsystems within organizations and in studying the environmental interfaces. But, it still has not advanced sufficiently to meet the needs. One of the major problems is that the practical need to deal with comprehensive systems of relationships is overrunning our ability to fully understand and predict these relationships. **We vitally need the systems paradigm but we are not sufficiently sophisticated to use it appropriately**. This is the dilemma. Do our current failures to fully utilize the systems paradigm suggest that we reject it and return to the older, more traditional, and time-tested paradigms? Or do we work with systems theory to make it more precise, to understand the relationships among subsystems, and to gather the informational inputs which are necessary to make the systems approach really work? We think the latter course offers the best opportunity.

Thus, we prefer to accept current limitations of systems theory, while working to reduce them and to develop more complete and sophisticated approaches for its application. We agree with Rapoport, who says:

The system approach to the study of man can be appreciated as an effort to restore meaning (in terms of intuitively grasped understanding of wholes) while adhering to the principles of **disciplined** generalizations and rigorous deduction. It is, in short, an attempt to make the study of man both scientific and meaningful.[111]

We are sympathetic with the second part of Rapoport's comment, the need to apply the systems approach but to make disciplined generalizations and rigorous deductions. This is a vital necessity and yet a major current limitation. We do have some indication that progress (although very slow) is being made.

WHAT DO WE NEED NOW?

Everything is related to everything else—but how? General systems theory provides us with the macro paradigm for the study of social organizations. As Scott and others

have pointed out, most sciences go through a macro-micro-macro cycle or sequence of emphasis[112]. Traditional bureaucratic theory provided the first major macro view of organizations. Administrative management theorists concentrated on the development of macro "principles of management" which were applicable to all organizations. When these macro views seemed incomplete (unable to explain important phenomena), attention turned to the micro level—more detailed analysis of components or parts of the organization, thus the interest in human relations, technology, or structural dimensions.

The systems approach returns us to the macro level with a new paradigm. General systems theory emphasizes a very high level of abstraction. Phillips classifies it as a third-order study[113] that attempts to develop macro concepts appropriate for all types of biological, physical, and social systems.

In our view, we are now ready to move down a level of abstraction to consider second-order systems studies or midrange concepts. These will be based on general systems theory but will be more concrete and will emphasize more specific characteristics and relationships in social organizations. They will operate within the broad paradigm of systems theory but at a less abstract level.

What should we call this new midrange level of analysis? Various authors have referred to it as a "contingency view," a study of "patterns of relationships," or a search for "configurations among subsystems." Lorsch and Lawrence reflect this view:

During the past few years there has been evident a new trend in the study of organizational phenomena. Underlying this new approach is the idea that the internal functioning of organizations must be consistent with the demands of the organization task, technology, or external environment, and the needs of its members if the organization is to be effective. Rather than searching for the panacea of the one best way to organize under all conditions, investigators have more and more tended to examine the functioning of organizations in relation to the needs of their particular members and the external pressures facing them. Basically, this approach seems to be leading to the development of a "contingency" theory of organization with the appropriate internal states and processes of the organization contingent upon external requirements and member needs.[114]

Numerous others have stressed a similar viewpoint. Thompson suggests that the essence of administration lies in understanding basic configurations which exist between the various subsystems and with the environment. "The basic function of administration appears to be coalignment, not merely of people (in coalitions) but of institutionalized action—of technology and task environment into a viable domain, and of organizational design and structure appropriate to it[115].

Bringing these ideas together we can provide a more precise definition of the contingency view. The contingency view of organizations and their management suggests that an organization is a system composed of subsystems and delineated by identifiable boundaries from its environmental suprasystem. The contingency view seeks to understand the interrelationships within and among subsystems as well as between the organization and its environment and to define patterns of relationships or configurations of variables. It emphasizes the multivariate nature of organizations and attempts to understand how organizations operate under varying conditions and in specific circumstances. Contingency views are ultimately directed toward suggesting organizational designs and managerial systems most appropriate for specific situations.

But, it is not enough to suggest that a "contingency view" is based on systems concepts of organizations and their management is more appropriate than the simplistic "principles approach." If organization theory is to advance and make contributions to managerial practice, it must define more explicitly certain patterns of relationships between organizational variables. This is the major challenge facing our field.

Just how do we go about using systems theory to develop these midrange or contingency views? We see no alternative but to engage in intensive comparative investigation of many organizations following the advice of Blau:

A theory of organization, whatever its specific nature, and regardless of how subtle the organizational processes it takes into account, has as its central aim to establish the constellations of characteristics that develop in organizations of various kinds. Comparative studies of many organizations are necessary, not alone to test the hypotheses implied by such a theory, but also to provide a basis for initial exploration and refinement of the theory by indicating the conditions on which relationships, originally assumed to hold universally are contingent. . . . Systematic research on many organizations that provides the data needed to determine the interrelationships between several organizational features is, however, extremely rare.[116]

Various conceptual designs for the comparative study of organizations and their subsystems are emerging to help in the development of a contingency view. We do not want to impose our model as to what should be considered in looking for these patterns of relationships. However, the tentative matrix shown in Exhibit 1-6 suggests this approach. We have used as a starting point the two polar organization types which have been emphasized in the literature— closed/stable/mechanistic and open/adaptive/organic.

We will consider the environmental suprasystem and organizational subsystems (goals and values, technical, structural, psychosocial, and managerial) plus various dimensions or characteristics of each of these systems. By way of illustration we have indicated several specific subcategories under the environmental suprasystem as well as the goals and values subsystem. This pro-

Exhibit 1-6. Matrix of Patterns of Relationships between Organization Types and Systems Variables

Organizational Supra-and Subsystems	Continuum of Organization Types	
	Closed/Stable/Mechanistic	Open/Adaptive/Organic
Environmental relationships		
General nature	Placid	Turbulent
Predictability	Certain, determinate	Uncertain, indeterminate
Boundary relationships	Relatively closed; limited to few participants (sales, purchasing, etc.); fixed and well-defined	Relatively open; many participant have external relationships; varied and not clearly defined
Goals and values		
Organizational goals in general	Efficient performance, stability, maintenance	Effective problem-solving, innovation, growth
Goal set	Single, clear-cut	Multiple, determined by necessity to satisfy a set of constraints
Stability	Stable	Unstable
Technical		
Structural		
Psychosocial		
Managerial		

cess would have to be completed and extended to all of the subsystems. The next step would be the development of appropriate descriptive language (based on research and conceptualization) for each relevant characteristic across the continuum of organization types. For example, on the "stability" dimension for goals and values we would have high, medium, and low at appropriate places on the continuum. If the entire matrix were filled in, it is likely that we could begin to discern patterns of relationships among subsystems.

We do not expect this matrix to provide **the** midrange model for everyone. It is highly doubtful that we will be able to follow through with the fieldwork investigations necessary to fill in all the squares. Nevertheless, it does illustrate a possible approach for the translation of more abstract general systems theory into an appropriate midrange model which is relevant for organization theory and management practice. Frankly, we see this as a major long-term effort on the part of many researchers, investigating a wide variety of organizations. In spite of the difficulties involved in such research, the endeavor has practical significance. Sophistication in the study of organizations will come when we have a more complete understanding of organizations as total systems (configurations or subsystems) so that we can prescribe more appropriate organizational designs and managerial systems. Ultimately, organization theory could serve as the foundation for more effective management practice.

APPLICATION OF SYSTEMS CONCEPTS TO MANAGEMENT PRACTICE

The study of organizations is an applied science because the resulting knowledge is relevant to problem-solving in ongoing institutions. Contributions to organization theory come from many sources. Deductive and inductive research in a variety of disciplines provide a theoretical base of propositions which are useful for understanding organizations and for managing them. Experience gained in management practice is also an important input to organization theory. In short, management is based on the body of knowledge generated by practical experience **and** eclectic scientific research concerning organizations. The body of knowledge developed through theory and research should be translatable into more effective organizational design and managerial practices.

Do systems concepts and contingency views provide a panacea for solving problems in organizations? The answer is an emphatic **no**; this approach does not provide "ten easy steps" to success in management. Such cookbook approaches, while seemingly applicable and easy to grasp, are usually shortsighted, narrow in perspective, and superficial—in short, unrealistic. Fundamental ideas, such as systems concepts and contingency views, are more difficult to comprehend. However, they facilitate more thorough understanding of complex situations and increase the likelihood of appropriate action.

It is important to recognize that many managers have used and will continue to use a systems approach and contingency views intuitively and implicitly. Without much knowledge of the underlying body of organization theory, they have an intuitive "sense of the situation," are flexible diagnosticians, and adjust their actions and decisions accordingly. Thus, systems concepts and contingency views are not new. However, if this approach to organization theory and management practice can be made more explicit, we can facilitate better management and more effective organizations.

Practicing managers in business firms, hospitals, and government agencies continue to function on a day-to-day basis. Therefore, they must use whatever theory is available, they cannot wait for the **ultimate** body of knowledge. (There is none!) Practitioners should be included in the search for new knowledge because they control access to a essential ingredient—organizational data—and they are the ones who ultimately put the theory to the test. Mutual understanding among managers, teachers, and researchers will facilitate the development of a relevant body of knowledge.

Simultaneously with the refinement of the body of knowledge, a concerted effort should be directed toward applying what we do know. We need ways of making systems and contingency views more usable. Without oversimplification, we need some relevant guidelines for practicing managers.

The general tenor of the contingency view is somewhere between simplistic, specific principles and complex, vague notions. It is a midrange concept which recognizes the complexity involved in managing modern organizations but uses patterns of relationships and/or configurations of subsystems in order to facilitate improved practice. The art of management depends on a reasonable success rate for actions in a probabilistic environment. Our hope is that systems concepts and contingency views, while continually being refined by scientists/researchers/theorists, will also be made more applicable.

REFERENCES AND NOTES

1. Bertalanffy, L. V. The theory of open systems in physics and biology. *Science*, 111:23–28, 1950.
2. Boulding, K. E. General system theory: the skeleton of science. *Management Sci.*, 2:197–208, 1956.
3. Wiener, N. *Cybernetics*. New York: John Wiley, 1948.
4. Kraegel, J. M. The Implementation of Federal Health Planning Policy through Area-Wide Health Systems Agencies. Ph.D. thesis, University of Wisconsin-Milwaukee, 1978.
5. Follett, M. P. Dynamic Administration. In *The Collected Papers of Mary Parker Follett*. New York: Harper & Brothers, 1940.
6. Taylor, F. *Scientific Management*. New York: Harper & Row, 1947.
7. Weber, M. *The Theory of Social and Economic Organization*. Henderson, A. M., and Parsons, T., (Trans.) and Parsons, T., (Ed.) New York: Oxford University Press, 1947.
8. McGuire, J. The Concept of the Firm. In Hill, W., and Egan, D. (Eds.) *Readings in Organizational Theory*. Boston: Allyn & Bacon, 1967.
9. McGuire, J. 1967.
10. Mayo, E. *The Human Problems of an Industrial Civilization*. New York: Macmillan, 1933.
11. Barnard, C. I. *The Functions of an Executive*. Cambridge, Mass.: Harvard University Press, 1938.
12. Haire, M. (Ed.) *Modern Organization Theory*. New York: John Wiley, 1959.
13. Scott, W. Organizational Theory: An Overview and an Appraisal. In Greenwood, W. (Ed.) *Management and Organizational Behavior Theories*. Cincinnati: Southwestern, 1965.
14. Baker, F. *Organizational Systems*. Homewood, Ill.: Richard D. Irwin, 1975.
15. Katz, D., and Kahn, R. L. *The Social Psychology of Organizations*. New York: John Wiley, 1966.
16. Georgopoulos, B. (Ed.) *Organization Research on Health Institutions*. Ann Arbor: University of Michigan, 1972.
17. Yuchtman, E. A Study of Organizational Effectiveness. Ph. D. thesis, University of Michigan, 1966.
18. Emery, P. E., and Trist, E. L. Socio-Technical Systems. In Churchman, C. W., and Verhulst, M. (Eds.) *Management Sciences: Models and Techniques*, vol. 2. London: Pergamon Press, 1960.
19. Thompson, J. *Organizations in Action: Social Science Basis of Administrative Theory*. New York: McGraw-Hill, 1967.
20. Lawrence, P., and Lorsch, J. *Organization and Environment*. Homewood, Ill.: Richard D. Irwin, 1969.
21. Bertalanffy, L. Von. *General Systems Theory*. New York: George Braziller, 1968.
22. Strother, B. Problems in the Development of a

Social Science of Organization. In Leavitt, H. (Ed.) *The Social Science of Organization*. Englewood Cliffs, N.J.: Prentice-Hall, 1963.
23. Thompson, J. 1967: Yuchtman, E. 1966.
24. Presthus, R. *Public Administration*. New York: Ronald, 1975. Easton, D. *A Systems Approach to Political Life*. Lafayette, Ind.: Purdue University Press, 1966.
25. Kaufman, H. Organization Theory and Political Theory. In Hill, W., and Egan, D. (Eds.) *Readings in Organization Theory*. Boston: Allyn & Bacon, 1967.
26. Aldrich, H. Organization Boundaries and Interorganizational Conflict. *Human Relations*. 24:279–93, 1971.
27. Selznick, P. *TVA and the Grass Roots*. Berkeley, Calif.: University of California Press, 1949.
28. Etzioni, A. Two approaches to organization analysis: a critique and a suggestion. *Admin. Sci. Quart.*, 5:257–78, 1960.
29. Yuchtman, E. 1967.
30. Thompson, J. 1967.
31. Gibson, J., Ivancevich, J., and Donnelly, J. *Organization, Structure, Processes, Behavior*. (rev. ed.) Dallas, Texas: Business Publications, 1976.
32. Bogdan, R., and Taylor, S. *Introduction to Qualitative Research Methods*. New York: John Wiley & Sons, Inc., 1975.
33. Gibson, J., Ivancevich, J. and Donnelly, J. 1976.
34. Bertalanffy, L. Von. *Perspectives on General Systems Theory*. Taschdjian, E. (Ed.) New York: George Braziller, 1975.
35. Evan, W. M. The Organization-Set: Toward a Theory of Interorganizational Relations. In Thompson, J. D. (Ed.) *Approaches to Organizational Design*. Pittsburgh: University of Pittsburgh Press, 1966.
36. Stogdill, R. Dimensions of Organizational Theory. In Thompson, J. D. (Ed.) *Approaches to Organizational Design*. Pittsburgh, Pa.: University of Pittsburgh Press, 1966, p.3.
37. Udy, S. H., Jr. *Organization of Work*. New Haven, Conn.: HRAF Press, 1959.
38. Parsons, T. *The Social System*. Glencoe, Ill: Free Press, 1957.
39. Udy, S. H., Jr. 1959.
40. Udy, S. H., Jr. 1959
41. Barnes, L. B. *Organizational Systems and Engineering Groups*. Boston: Harvard Business School, Division of Research, 1960.
42. Blau, P. M., and Scott, W. R. *Formal Organizations*. San Francisco: Chandler, 1962.
43. Form, H. W., and Miller, D. C. *Industry, Labor, and Community*. New York: Harper & Row, 1960.
44. Riley, J. W. *The Corporation and Its Publics*. New York: Wiley, 1963.
45. Rice, A. K. *The Enterprise and Its Environments*.

London: Tavistock, 1963. Thompson, J. D., and Bates, F. L. Technology, Organization and Administration. In Thompson, J. D. et al. (Eds.) *Comparative Studies in Administration*. Pittsburgh, Pa.: University of Pittsburgh Press, 1959.

46. Ginzberg, E., and Reilley, E. W. *Effecting Change in Large Organizations*. New York: Columbia University Press, 1957.

47. Ronken, H. O., and Lawrence, P. R. *Administering Changes*. Boston: Harvard University Graduate School of Business Administration, 1952.

48. Jaques, E. *The Changing Culture of a Factory*. New York: Dryden, 1952.

49. Lawrence, P. R. *Organizational Behavior and Administration*. Homewood, Ill.: Dorsey, 1961.

50. Bennis, W. G., Benne, K. D., and Chin, R. *The Planning of Change*. New York: Holt Rinehart and Winston, 1961.

51. Baum, B. H. *Decentralization of Authority in a Bureaucracy*. Englewood Cliffs, N.J.: Prentice-Hall, 1961.

52. Blau, P. M. *Bureaucracy in Modern Society*. New York: Random House, 1956.

53. Gouldner, A. W. *Patterns of Industrial Bureaucracy*. Glencoe, Ill.: Free Press, 1954.

54. Child, J. Organizational Design and Performance: Contingency Theory and Beyond. In Burack, E. H., and Negandhi, A. (Eds.) *Organization Design: Theoretical Perspectives and Empirical Findings*. Kent, Ohio: Kent State University Press, 1977.

55. Baker, F. The Changing Hospital Organizational System: A Model for Evaluation. In Rubin, M. D. (Ed.) *Man in Systems*. New York: Gordon and Breach, 1971.

56. Schulberg, H. C., Caplan, G., and Greenblatt, M. Evaluating the changing mental hospital: a suggested research strategy. *Mental Hygiene*, 52:218–25, 1968. Schulberg, H. C., and Baker, F. The Changing Mental Hospital: Is It Really Changing? Paper Presented at the 20th Mental Hospital Institute, Washington, D.C., Oct. 1, 1968.

57. Scott, W. G. Organization theory: an overview and appraisal. *J. Acad. Management*, 4:7–26, 1961.

58. Katz, D., and Kahn, R. L. 1966.

59. Etzioni, A. 1960. Levine, S., and White, P. Exchange as a conceptual framework for the study of interorganizational relationships. *Admin. Sci. Quart.*, 5:583–601, 1961. Emery, P. E., and Trist, E. L. 1960.

60. Bertalanffy, L. Von. 1950.

61. Emery, P. E., and Trist, E. L. 1960. Miller, E. J., and Rice, A. K. *Systems of Organization*. London: Tavistock, 1967.

62. Rice, A. K. *The Enterprise and the Environment*. London: Tavistock, 1963.

63. Sofer, C. *The Organization from within*. Chicago: Quadrangle, 1961.

64. Etzioni, A. 1960, p. 260.

65. Baker, F., and Schulberg, H. C. The development of a community mental health ideology scale. *Community Mental Health J.*, 3:216–25, 1967.

66. Caplan, G. *An Approach to Community Mental Health*. New York: Grune and Stratton, 1961.

67. Gilbert D., and Levinson, D. J. Role Performance Ideology and Personality in Mental Hospital Aides. In Greenblatt, M., Levinson, D.J., and Williams, R. H. (Eds.) *The Patient and the Mental Hospital*. New York: Free Press, 1957, pp. 197–208. Sharaf, M. R., and Levinson, D. J. Patterns of Ideology and Role Definition among Psychiatric Residents. In Greenblatt, M., Levinson, D.J., and Williams, R.H. (Eds.) 1957, pp. 263–85.

68. Rice, A.K. 1963, p. 18.

69. Bertalanffy, L. Von. 1950, p. 25.

70. Thompson, J.D., and McEwen, W.J. Organizational Goals and Environment: Goal Setting as an Inter-Action Process. *Am. Sociological Rev.*, 23:23–31, 1958.

71. Levine, S. and White, P.E., Exchange as a Conceptual Framework for the Study of Interorganizational Relationships. *Admin. Sci. Quart.*, 5:583–601, 1961.

72. Baker, F. 1971, p. 325.

73. Homans, G. C. Social behavior as exchange. *Am. J. Sociology*, 8:597–606, 1958.

74. Weber states that "by 'exchange' in the broadest sense will be meant every case of a formally voluntary agreement involving the offer of any sort of present, continuing, or future utility in exchange for utilities of any sort offered in return." Weber employs the term "utility" in the economic sense. It is the "utility" of the "object of exchange" to the parties concerned that produces exchange. See Weber, M. *The Theory of Social and Economic Organization*. New York, 1947, p. 170. Homans, on the other hand, in characterizing interaction between persons as an exchange of goods, material and nonmaterial, sees the impulse to "exchange" in the psychological make-up of the parties to the exchange. He states, "the paradigm of elementary social behavior, and the problem of the elementary sociologist is to state propositions relating the variations in the values and costs of each man to his frequency distribution of behavior among alternatives, where the values (in the mathematical sense) taken by these variables for one man determine in part their values for the other." See Homans, G. C. 1958, p. 598.

75. Parsons, T. Suggestions for a sociological approach to the theory of organizations—I. *Admin. Sci. Quart.*, 1:63–85, 1956.

76. Parsons, T. 1956, p. 64.

77. Gouldner, A. W. Reciprocity and Autonomy in Functional Theory. In Gross, L. (Ed.) *Symposium on Sociological Theory*. Evanston, Ill.: 1959. Also, the norm of reciprocity: a preliminary state-

ment. *Am. Sociological Rev.*, 25:161–78, 1960.
78. Sills, D. L. *The Volunteers: Means and Ends in a National Organization.* Glencoe, Ill.: Free Press, 1957.
79. In our research a large percentage of our respondents spontaneously referred to the undesirability of overlapping or duplicated services.
80. Yuchtman, E., and Seashore, S. E. A system resource approach to organization effectiveness. *Am. Sociological Rev.*, 32:891–903, 1967.
81. Gamson, W. A. *Power and Discontent.* Homewood, Ill: Dorsey, 1968.
82. Presthus, R. 1975.
83. Yuchtman, E., and Seashore, S.E. 1967, p. 499.
84. Thompson, J. 1967.
85. An entire article could be devoted to a discussion of ingenious ways in which the term "systems approach" has been used in the literature pertinent to organization theory and management practice.
86. Deutsh, K. W. Toward a Cybernetic Model of Man and Society. In Buckley, W. (Ed.) *Modern Systems Research for the Behavioral Scientist.* Chicago: Aldine, 1968, p. 389.
87. Phillips, D. C. Systems Theory—A Discredited Philosophy. In Schoderbek, P. P. (Ed.) *Management Systems.* New York: John Wiley, 1971, p. 56.
88. Barnard, C. I. 1938, p. 65.
89. Miller, R. F. The new science of administration in the USSR. *Admin. Sci. Quart.*, Sept. 1971.
90. Scott, W. G. 1961.
91. Simin, H. A. The Architecture of Complexity. In Litterer, J.A. (Ed.) *Organizations: Systems, Control and Adaptation*, vol. 2. New York: John Wiley, 1969.
92. Boulding, K. E. 1956. Buckley, W. 1968. Easton, D. *A Systems Analysis of Political Life.* New York: John Wiley, 1965. Hall, A. D., and Eagen, R. E. Definition of System. *General Systems, Yearbook for the Society for the Advancement of General Systems Theory*, vol. 1, 1956. Miller, J. G. Living systems: basic concepts. *Behavioral Sci.*, July 1965. Parson, S. T. *The Social System.* Glencoe, Ill.: Free Press, 1951. Thompson, J. D. 1967.
93. Churchman, C. W. *The Systems Approach.* New York: Dell, 1968. Emery, F. E., and Trist, E. L. 1960. Kast, F. E., and Rosenzweig, J. E. *Organization and Management Theory: A Systems Approach.* New York: McGraw-Hill, 1970. Katz, D., and Kahn, R. L. 1966. Litterer, J. A. *Organizations: Structure and Behavior*, vols. 1, 2.

New York: John Wiley, 1969. Schein, E. *Organizational Psychology*, rev. ed. Englewood Cliffs, N.J.: Prentice-Hall, 1970.
94. Miller, J. G. 1965.
95. Kast, F., and Rosenzweig, J. E. 1970. Katz, D., and Kahn, R. L. 1966. Litterer, J. A. 1969. Rice, A. K. *The Modern University*, London: Tavistock, 1970. Scott, W. G. 1961.
96. Back, K. W. Biological models of social change. *Am. Sociological Rev.*, Aug. 1971, p. 660.
97. Katz, D., and Kahn, R. L. 1966, p. 31.
98. Silverman, D. *The Theory of Organizations.* New York: Basic Books, 1971, p. 31.
99. Rapport, A., and Horvath, W. J. *Thoughts on Organization Theory.* In Buckley, W. 1968, pp. 74–75.
100. Ackoff, R. L. Towards a system of systems concepts. *Management Sci.*, July 1971, p. 669.
101. Murray, H. A. Preparation for the Scaffold of a Comprehensive System. In Koch, S. (Ed.) *Psychology: A Study of a Science*, vol. 3. New York: McGraw-Hill, 1959, pp. 50–51.
102. Miller, J. G. 1965, p. 216.
103. Katz, D., and Kahn, R. L. 1966, p. 33.
104. Yuchtman, E., and Seashore, S. E. 1967, p. 896.
105. Yuchtman, E., and Seashore, S. E. 1967, p. 891.
106. Kuhn, T. S. *The Structure of Scientific Revolutions.* Chicago: University of Chicago Press, 1962.
107. Kuhn, T. S. 1962, p. 110.
108. Kuhn, T. S. 1962, p. 23.
109. Buckley, W. 1968. Easton, D. 1965. Katz, D., and Kahn, R. L. 1966. Litterer, J. A. 1969. Miller, E. J., and Rice, A. K. 1967. Rice, A. K. 1970. Thompson, J. D. 1967.
110. Burns, T., and Stalker, G. M. *The Management of Innovation.* London: Tavistock, 1961.
111. Miller, E. J., and Rice, A. K. 1967.
112. Chamberlain, N. W. *Enterprise and Environment: The Firm in Time and Place.* New York: McGraw-Hill, 1968.
113. Buckley, W., 1968, p. xxii.
114. Scott, W. G. 1961.
115. Phillips, D. C. 1971.
116. Lorsch, J. W., and Lawrence, P. R. *Studies in Organizational Design.* Homewood, Ill.: Irwin-Dorsey, 1970, p. 1.
117. Thompson, J. D., 1967, p. 157.
118. Blau, P. M. The comparative study of organizations. *Industrial and Labor Relations Rev.*, April 1965.

UNIT II   ENVIRONMENTS

- **THE TASK ENVIRONMENT**
- **CONTINGENCY THEORY**
- **THE SOCIOCULTURAL ENVIRONMENT**
- **THE POLITICAL ENVIRONMENT**

Environment has been defined as that portion outside of the system which can be observed to interact with the system (see Glossary). Boundaries separate a system from its environment. The environment of a system consists of those elements and interactions among elements outside the system boundary that affect the system and are affected by the system[1]. The environment of an organization as a system is composed of a large number of elements and interactions outside of the organization's boundaries. Various authors have attempted to categorize these in a meaningful way. One categorization is the continuum proposed by William Evans, who suggests that confusion surrounds the concept of environment because each theory deals with a different point on a micro-macro continuum. At the micro end of the continuum, the focus is on the internal environment of the organization where individuals act out their roles. At the macro end of the continuum, the organization is the unit of analysis, and environmental elements include generalized external environmental influences such as the state of technology in a society. Evans[2] adds that the proliferation of terms relating to different dimensions of the environment impinging on organizations is a reflection of the growing pains of organization theory as it becomes macro-oriented.

The second schema of environmental elements offered here is that of Osborn and Hunt[3]. They state that while the environment of an organization is composed of an infinite set of elements outside the boundaries of the organization, the most important segments of these elements are the other organizations, associations of individuals, and broad forces influencing the organization. They group these into three categories: macro, aggregation, and task environment.

*The **macro** environment is the general cultural context of a*

75

*specific geographical area and contains recognized and important influences on organizational characteristics and outputs. Richman and Farmer[4] label these forces economic, educational, legal-political, and social-cultural. The **aggregation** environment consists of the associations, interest groups, and constituencies operating within a macro environment. The **task** environment is defined by Osborn and Hunt as that portion of the total setting which is relevant for goal setting and goal attainment of the organization. The macro and aggregation segments are the larger, more general framework within which all organizations in a state, nation, or geographic area must operate. The task environment consists of all organizations with which a given organization must interact to grow and survive.*

The third conceptualization of environment to be described is that of Van de Ven, Emmett, and Koenig[5], who recognize three approaches to the study of organization environments: 1) the environment as an external constraining phenomenon, 2) the environment as a collection of interacting groups or persons, and 3) the environment as a social system.

External constraining phenomena include both general influences (e.g., legal, political, ecological) and specific influences (the task environment). Studies of "collections of interacting groups and persons" form a relatively new body of knowledge called interorganization theory (diagram 2 in Exhibit 2-1). In this theory, the focus is upon the network of exchange relationships between organizations. The "social system" view (diagram 3 in Exhibit 2-1c) is derived from the theory of Talcott Parsons. The actions of the organizations are conceptualized as interdependent; over time, member organizations take on specialized roles and develop behavioral expectations of each other regarding the rights and obligations of membership. Studies which employ the social system approach are in a rudimentary stage of development.

Many other environmental classifications exist in the research literature. Though their areas of emphasis differ, there is considerable similarity in the environmental variables, identified as influencing organizational behavior.

All of us are vaguely aware that availability of resources in an environment affects an organization's ability to function. Scholars who have conducted the more recent research on organization-environment relationships provide evidence of the extent to which groups and forces in the environment cause variations in an organization's structure and processes, as well as in its effectiveness.

The first studies of organizational environment focused on the task environment. Later studies moved away from the focus on

Exhibit 2-1. Conceptualizations of Organization/Environment Relationships

a. EVANS

MICRO   MACRO

INTERNAL   EXTERNAL
ENVIRONMENT   GENERALIZED
of   INFLUENCES
ORGANIZATION

b. OSBORN & HUNT

TASK ENVIRONMENT	AGGREGATION	MACRO ENVIRONMENT
All organizations which are necessary for survival and growth	Associations Interest groups Constituencies	Economic Educational Legal, Political Socio-cultural

c. VAN de VEN, EMMETT, and KOENIG

1. Constraining Environment	2. Collection of Interacting Groups	3. Social System

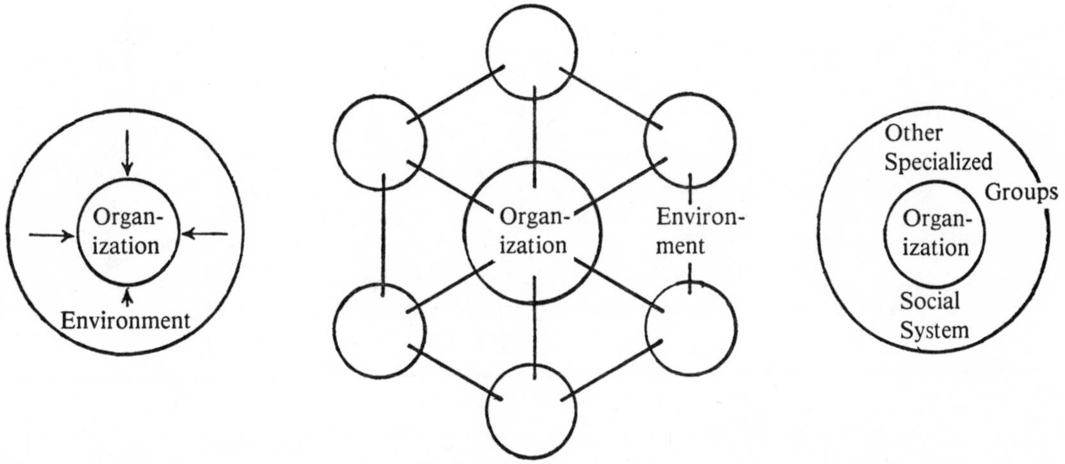

task environment, the micro end of Evans' conceptual continuum, toward the more expansive macro environment. Macroenvironmental variables are of a more generalized nature and difficult to operationalize or isolate; thus, the later studies took a much different form than the early task environment research.

Three macroenvironmental studies conducted in the health field are included in this unit to illustrate how social, political, and economic factors changed the structure and behavior of the service agencies within specific geographic areas. In John R. Kimberly's[6] study of 123 rehabilitation organizations, the results suggest that the rehabilitation agencies developed structures and programs with increased visibility in order to attract essential resources from individuals who controlled those resources. Daniel M. Harris demonstrates in his paper that hospital utilization is a social process embedded in a web of social causation. The third study, which was conducted in a Chicano community of Chicago, demonstrates how change in a community mental health center was a direct result of socioeconomic factors of the environment. Community collaboration took place in order to adapt the system to fit the needs of the particular ethnic community. The selections in this unit address widely varying situations. All of them indicate that organizations need to shape health care delivery patterns in response to the environmental circumstances pressing upon them.

6. The Task Environment

One of the first studies of the task environment was William Dill's 1958 study of the task environment of two Norwegian manufacturing firms. Dill used the data he collected to identify the impact of environmental factors on behavior in organizations[7]. He conceptualized the environment as a flow of information to participants in an organization, compared the accessibility of the information to the two different managements, and reported on how those variables influenced the structuring of management and the autonomous behavior of the managers in the two firms. He emphasized the fact that the cognitive activities of the organization managers serve to link environment stimuli (information) and the managers' behavioral responses. The two firms differed sharply in the degree to which the department heads and staff thought and behaved autonomously, both with respect to their peers and with respect to the top management of the firms.

Because Dill's definition of task environment was of such importance that it continues to be cited frequently in organizational literature, it is reprinted here with Dill's discussion of it:

Each management group planned action on the basis of information it received about environmental events. I have denoted that part of the total environment of management which was potentially relevant to goal setting and goal attainment as the task environment.

The task environment of management consisted of inputs of information from external sources. These inputs did not represent "tasks" for the organization; by task I mean a cognitive formulation consisting of a goal and usually also of constraints on behaviors appropriate for reaching the goal. When we study the task environment, we are focusing on the stimuli to which an organization is exposed; but when we study tasks, we are studying the organization's interpretations of what environmental inputs mean for behavior. These interpretations are subject to errors of perception and to the bias of past experience.

The task environment, as information inputs, and tasks, as cognitive formula-

tions to guide action, need further to be distinguished from task-fulfilling activities, the actual behavior of men in organizations. In many studies where task variables have been considered, clear distinctions have not been made among things that the organization does (activities), things that the organization sets itself to do (tasks), or stimuli that the organization might respond to (task environment). There are many relevant inputs of information which organizations do not attend to as well as many tasks which they formulate but never act upon.[8]

*For both of the Norwegian firms Dill studied, the elements of task environment that had the greatest impact on the organization's goal attainment included **customers** (distributors and users), **suppliers** (of materials, labor, equipment, capital, and work space), **competitors** (for both markets and resources), and **regulator groups** (government agencies, unions, and interfirm associations). The external environmental factors Dill identified as relevant included the degree of its **homogeneity** (the assortment of unions, markets, competitors to be dealt with), the **stability** (in market size, technology, new products), **disruptiveness of environmental inputs, demands for personal interaction** by environmental groups, **the routing of inputs** (methods of transmitting information), and **complexity of inputs** (skills required to process transactions).*

Most analysts of task environment place an emphasis on the organization and groups (such as consumers) with which the system under analysis interacts, although occasionally general economic and social variables are addressed. Dill's description of task environment gives primary significance to four groups: customers, suppliers, competitors, and regulating groups although he also included environment characteristics such as homogeneity. Employees were not included as task environment members by Dill because they are considered by him to be a subsystem of the organization, not a part of the external environment. Osborn and Hunt[9] differ from Dill in their definition of task environment. They offer strong arguments for restricting task environments to organizations, thus differentiating the effects of environmental conditions from the effects of interactions between the system and groups in the external environment.

The genius of Dill's research was the intuitive way he examined the interplay of variables lying within and outside the boundaries of the system. He used the managers of the organization as the focus of analysis rather than the larger organization itself. In the more recent studies of organization-environment relationships, researchers focus on the organization as a distinctive entity and devote little attention to the individuals working within it. These later studies are at the macro end of William Evans's micro-macro continuum.

Dill's research perspective was unique and had considerable impact on the closed-system thinking of that time. He described how the primary influences—customers, suppliers, competitors, and regulatory groups—as well as generalized influences such as the short supply of skilled workers or political conditions affect the organization's performance. Today, we are much better oriented towards recognition of the effect of external environmental factors on organizational goals. A case in point is the current effort of hospitals to control hospital costs, a commendable goal, which David Kinzer[10] argues is hampered by the hospital's inability to control its customers (patients), suppliers (physicians), or the regulatory group (health systems agencies). He describes the combined impact on these external groups on a hospital's performance as follows:

Everyone knows that what a doctor decides his patient needs is the most "cost consequential" of all decisions inside the hospital . . . The problem of controlling hospital costs comes into sharp focus here. Nobody (conspicuously including the hospital administrator) is really telling the doctor what to decide. The control systems stop short of that.

Hospitals are penalized for substandard occupancy, but they can't prevent too many of their surgeons from taking simultaneous Florida vacations. The mix of hospital admissions is controlled by the presence (or absence) of admitting physicians, and which specialists happen to be busiest (or most in vogue) and by other external and uncontrollable forces. Most hospitals strive to get more superspecialists to help build their census, but they lose here, too, if these specialists admit too many high-intensity cases.

Our hospitals are now being told to provide services that health planners say are needed by the people in the prescribed service areas, when these may have little to do with what the patients admitted by their medical staffs need. The fact that graduates of the same medical school tend to trade-off patients may have more relevance to hospital planning than any elaborate formula developed to date under our CoN (Certificate of Need) program.[11]

The hospital performance goal of cost containment, of course, is a response to strong external political pressure exerted by the current federal legislature and administration. Given the political context in which today's hospitals function and their lack of control over their suppliers of patients, they have great difficulty achieving their organizational goals of solvency and cost containment.

7. Contingency Theory

*Richman and Farmer[12] state in **Management and Organizations** that if the external environment directly and significantly influences organization performance, and if the critical environment factors can be identified and isolated, it should be possible to alter or react to the environment to improve efficiency and effectiveness. Even though an organization functions in any environment containing many constraints, its management need not be passive victims of the environment. Effective managers can usually work around environmental constraints and even create opportunities for the organization because of them. Richman and Farmer's approach to the problem was to develop a list of critical factors which generally constrain or help an enterprise. They suggest that research determine how each affects the organization in various situations.*

Open-system organization research of the last 15 years has attempted to describe how environmental constraints, organizational structure, and leadership are linked to performance of the organization and satisfaction of its members. Dominant in this field of organization research is contingency theory. Contingency theory suggests that the internal structure of an organization must be consistent with the requirements of the tasks that an organization is performing; i.e., there must be a "fit" between internal organizational characteristics and external environmental requirements if an organization is to perform effectively in dealing with its environment. Lorsch[13] states there are two related aspects to the fit: 1) each fundamental unit (sales, production, research, and so on) must have internal characteristics consistent with the demands of its particular sector of the total environment, and 2) the total organization must achieve, in spite of the differentiation among its units, the pattern of integration required by the total environment. Lorsch's research, for instance, suggests that highly uncertain environments require tight organization for effectiveness. Lorsch raises an interesting question. Does the match of organization unit and environment simply meet information processing requirements of the environment, or does this consistency also have an effect on individual organization members which motivates them towards effective performance? Lorsch asks for more research on the terms of the exchange relationship between organizational, individual, and environmental factors. "By focusing on the relationship among organizational, individual, and environmental factors, we hope to build a bridge between the concerns of macro-organization theorists with organizational and environmental variables, and the focus of psychologists on the individual as the unit of analysis," he states[14].

Contingency theory is based on the belief that organizations (unlike biological systems) have the ability to modify themselves, and do so as a desired self-correcting strategy which gives them control over external environmental demands.

Major contingencies studied by contingency theorists include the characteristics and diversity of environments in which the organization operates, the technology the organization utilizes, its scale of operations, and the type of people it employs. John Child comments on the contribution of contingency theory to our understanding of organization-environment relationships as follows:

No organization operates in a vacuum. Certain elements in its context, such as type of environment or scale of operation, will probably have been deliberately selected as part of its strategy. Some environments may be more beneficial and provide greater opportunities than others; similarly, there may be economies of scale open to the larger organization. While, however, these strategic choices create the conditions for improved performance, an organization may not be able to capitalize on them fully unless it devises a set of administrative arrangements adapted to the operations required by its environment, to the complexity and scale of its tasks, to the expectations and needs of the people performing these tasks, and to the control and integration requirements stemming from these factors. The design of administration and structure is seen in this way to modify the direct effect of contextual factors upon performance.

The environment has normally been regarded as the most significant contingency for organizational design following pioneering researches of Burns and Stalker (1961)[15] and Lawrence and Lorsch (1967)[16]. According to contingency theory, the degree of variability and of complexity characterizing an organization's environment will indicate which approach to organizational design is appropriate. Variability refers to the presence of changes which are relatively difficult to predict, involve important departures from previous conditions, and are likely, therefore, to generate considerable uncertainty. Complexity of the environment is said to be greater, the more extensive and differentiated are the external sectors in which the organization operates.[17]

Contingency factors of the external environment as described by Child thus include environmental complexity, technology (variability of inputs), the competitive situation, and environmental variability.

The competitive situation is sometimes referred to as the variable of organization dependence. Child states that it is coming to be recognized as a major explanatory factor for an organization's structural and performance variation. An organization which has achieved some degree of monopoly or has found a protected niche in the environment is often able to control or ignore environmental contingencies.

Contingency theorists have been accused by the more tradition-bound theorists of having an eclectic approach, of not operationalizing their definitions adequately, of failing to recognize the possibility of multiple contingencies, of using too narrow a definition of external environment[18]. Even if there is truth in all of these criticisms, the fact remains that through these scholars' efforts, impressive progress has been made in identifying, conceptualizing, and operationalizing relevant environmental variables affecting organization structure and effectiveness.

ENRIVONMENTAL CONSTRAINTS AND ORGANIZATIONAL STRUCTURE: A COMPARATIVE ANALYSIS OF REHABILITATION ORGANIZATIONS

By John R. Kimberly

Reprinted with permission from *Administrative Science Quarterly*, Volume 20, Number 1, March 1975, pp. 1-9.

Research such as the following study of rehabilitation organizations builds on the contingency theorists' pioneering efforts to identify organization-environment relationships.

This study examines changes in dominant values and levels of technological development, conceptualized as environmental constraints, and relates them to differences in orientation and structure in a sample of rehabilitation organizations. The relationship found between date of organizational founding and orientation is interpreted as tentative support for propositions about the general relationship between social structure and organizational structure. The relationship found between income from grants and orientation indicates the importance of external control of resources as one type of environmental constraint[19].

The sociological literature provides many examples of attempts to establish links, both implicit and explicit, between social structure and organizations. Bendix[20], for example, explored the relationship between dominant political ideology and how authority of managers over subordinates was legitimated in an industrial context. Abegglen[21] found that certain features of the Japanese social structure were reflected in the social organization of the factory. And Crozier[22] examined how certain characteristics of French society were embedded in the French bureaucratic system. These examples demonstrate that there are systematic relationships between social structure and organizations; they do not, however, provide any evidence about how changes in organizational structure might be related to changes in social structure.

THEORETICAL BACKGROUND

The view developed in this paper owes a great deal to two earlier formulations. Stinchcombe, in focusing on one aspect of the general relationship posited here, stated that:

The organizational inventions that can be made at a particular time in history depend on the social technology available at the time. Organizations which have purposes that can be efficiently reached with the socially possible organizational forms tend to be founded during the period in which they become possible.[23]

This argument has recently been more fully developed and is discussed at some length by Hall[24].

87

Burns, in a discussion of the comparative study of organizations stated that:

Variation and developmental change in the value system of society, as in the educational system and the technological system, manifestly affect profoundly not only the institutional pattern and structure of organizations, not only the capacity of society to meet demand and use resources, but the kinds of organizations which may be brought into being. . . . [25]

If these two statements are combined with system-theoretic approaches to organizational analysis, the problem can be conceptualized as follows: As open systems, organizations engage in various transactions with their environments. These transactions are complex, variable across organizations and environments, and reciprocal. At a given time, however, there are various environmental constraints which limit the structural form that organizations can adopt. Thus, the etiology of organizational configurations is, at least in part, a function of environmental influences, and variability in these configurations should be predictably related to variability in environmental influences.

Clearly, specification of the influences relevant to one organization or set of organizations, say prisons, may not be relevant for another, say manufacturing companies. However, examination of the general proposition in the light of data from one particular set of organizations may provide a basis for attempts to use the approach in the analysis of other sets.

Sheltered Workshops

The particular type of rehabilitation organization studied here, the sheltered workshop, is defined as "a work-oriented rehabilitation facility with a controlled working environment and individual vocational goals which utilizes work experience

and related service for assisting the handicapped person to progress toward normal living and productive vocational status"[26]. The rehabilitation activities of sheltered workshops, therefore, have two main objectives—preparing the client 1) to cope with the physiological requirements of work and 2) to meet the psychological and social demands of work.

Besides providing rehabilitation, however, the workshop must also maintain itself economically. It produces goods or provides services for its community, and the income from this production provides remuneration for its clients and contributes toward maintaining its economic viability. The workshop must be concerned, therefore, with many of the problems of any small business organization.

Although all workshops engage in these two chief activities—rehabilitation and production—they do not all place equal emphasis on production activities and rehabilitation activities. Field observation in a large number of workshops suggested that different operational strategies were used by different workshops, and that these differences were reflected in the proportion of total income derived from goods-producing activities. Where a relatively high proportion of the total income of the workshop was derived from production activities, clients were viewed as employees and thus as instruments of production; where a relatively low proportion of the total income was derived from production activities, clients were viewed as individuals needing various forms of therapy to enhance their life opportunities. Although this statement oversimplifies the observations made, it does characterize the two polar types of strategy observed.

Based on these observations, a more systematic analysis was made of the extent to which differences in other organizational characteristics were related to differences in emphasis on production activities, or orientation. Workshops receiving half or more of their income from production activities, that is, production-oriented workshops, had significantly different structural characteristics from workshops, with less than half of their income from this source, that is, rehabilitation-oriented workshops. In an analysis reported elsewhere[27], it was found that workshops with a rehabilitation orientation had lower levels of technological complexity, higher numbers of services and programs, higher professionalization of staff, higher proportions of severely impaired clients, higher rates of rehabilitation success, proportionately fewer members of their boards of directors from industry, and were located more often in communities with highly developed health systems than workshops with a production orientation. All of these differences were statistically significant. The theoretical perspective developed in this paper would suggest that these differences might be explained, at least in part, on the basis of differences in environmental constraints existing at the time the organizations were founded.

History of Environmental Constraints on Workshops

Identification of the external factors seen as relevant for the organizations studied was based on an examination of the history of the workshop movement. Since there was no comprehensive history of sheltered workshops, it was necessary to draw on information from a wide range of sources. Based on scattered published materials and consultation with various experts in the field, the history of the movement was divided into two periods.

Although the first sheltered workshop was established before the Civil War, the development of workshops was insignificant until the early twentieth century[28]. Early concern was centered on handicapped children, particularly crippled children; there were few organized attempts to provide aid to handicapped adults before World War I. Services were primarily custodial and tended to isolate the invalid from the existing medical technology[29]; the concept of rehabilitation was not yet part of the thinking of social planners or medical specialists.

"The unprecedented numbers of casualties incurred in World War I, together with the ability of medical science to keep alive many of the maimed who in former wars would certainly have died . . . [30] clearly gave impetus to the rehabilitation movement. The nation's ideological concern to try to repay those who had sacrificed their physiological well-being in defense of the country encouraged the development of organizations to care for the handicapped. A special division for the physical reconstruction of the wounded was established in the Office of the Surgeon General of the Army in August 1917, with subdivisions of education and physiotherapy. The Smith-Sears Vocational Rehabilitation Act of 1918 placed discharged veterans under the direction of the Federal Board for Vocational Education. Subsequent legislation in 1920 broadened eligibility to include disabled civilians[31]. The growing numbers of victims of industrial accidents during this period provided further impetus to the development of sheltered workshops in some of the larger cities[32].

Later, the social philosophy of the New Deal provided a favorable climate. The Wagner-O'Day Act of 1938 provided for government purchase of certain articles manufactured in workshops for the blind and established National Industries for the Blind, a central nonprofit agency to facilitate the distribution of orders among agencies for the blind. The Fair Labor Standards Act allowed for special certificates for payment of wages lower than the minimum wage rate to handicapped persons producing goods for interstate commerce and for time-and-a-half pay for all work done in excess of the normal 40-hour week[33].

The end of World War II was the end of this first period and the beginning of a second period. It is generally agreed that rehabilitation as a social movement took its greatest strides in the years during and immediately following World War II, and that the number of sheltered workshops increased as a result of the increased demands for rehabilitation services. As Wessen noted:

It was during World War II, with its new and overwhelming challenges to medicine and to social responsibility for the handicapped, that provided the impetus for rehabilitation to develop to its present state. During this period—and particularly in military hospitals—new philosophies of patient care were evolved which greatly increased the potential for rehabilitation of the handicapped.[34]

And, with special relevance to sheltered workshops:

Since World War II, the sheltered workshop has emerged as a strong and unique element in the rapidly expanding network of specialized rehabilitation services. There has been a slow but steady movement away from the early concept of the workshop as a custodial care institution and a recognition that the proper workshop objective is the preparation of disabled individuals for competitive employment and a regular earned wage. Emphasis is upon the value of work to the individual and the dignity which comes from social usefulness and economic independence.[35]

The Barden-LaFollette Act, passed during the Second World War, included physical restoration within the scope of government-sponsored vocational rehabilitation activities, providing handicapped clients with physical examinations, treatment, and hospitalization for psychiatric as well as physical disorders at government expense.

"By 1954, it was evident that rehabilitation as a social movement had developed the techniques and the institutional patterns which, if broadly implemented, could promise massive achievements for the nation's disabled. However, funds, personnel and facilities were lacking."[36]. To help remedy this situation Congress passed amendments to Public Law 565, the Vocational Rehabilitation Act of 1954, increasing the numbers of clients served by the vocational rehabilitation agencies by more than 300 percent within five years, and offering grant programs to help voluntary organizations in rehabilitation to expand and improve services, to conduct research in rehabilitation, and to enlarge workshops and facilities[37].

Hypothesis

Following our basic idea that factors external to the organization act as a constraint on its structure, one would expect that workshops founded before and during World War II would be more likely to be production-oriented and those founded after World War II would be more likely to be rehabilitation-oriented.

DATA ANALYSIS AND RESULTS

Sample

The sample consisted of 123 sheltered workshops located in New York, New Jersey, and Pennsylvania. The data used were gathered through a questionnaire sponsored jointly by the Region II Rehabilitation Research Institute at Cornell University and the state rehabilitation agencies in 1966. Of the 171 workshops surveyed, 86 percent responded to this questionnaire; but, because of incomplete data and the fact that 10 were in operation less than a year, only 76 percent of those surveyed were included in the final sample. These 123 organizations served 91 percent of the clients receiving services in this region. The dividing line for the earlier and later period of workshops was 1946. The orientation measure was based on the proportion of total income derived from production activities: organizations with half or more of their income from production activities were classified as production-oriented; those with less than half of their income from this source were classified as rehabilitation-oriented.

Data and Analysis

To test the hypothesis, various analyses were undertaken. Only 6 of the 33 workshops founded before 1946 were rehabilitation-oriented, while 58 of the 90 workshops founded after 1946 were rehabilitation-oriented[38].

On the basis of these results, the null hypothesis that there is no relationship between period of organizational foundation and orientation was rejected. They suggest that, given the relationships between orientation and other organizational characteristics previously reported, differences in organizational characteristics are not independent of differences in environmental constraints, and they offer empirical support for the specific hypothesis tested. The results of the statistical techniques used to test the hypothesis were consistently significant and in the predicted direction. Because the sample from which these data were collected was not a probability sample, however, interpretation of levels of significance is problematic.

The figure displays graphically the relationship between date of founding and orientation and highlights an important trend in the data. While more production-oriented workshops were actually founded after the War than before or during it, in the period after the War they were founded at a decreasing rate. If this trend continued beyond 1966, it would lend further support to the hypothesis.

In considering these data, it should be pointed out that they were collected about the operations of organizations in 1966. This raises the question of what effects, if any, organizational growth and development might have had on the measure of orientation. Because no longitudinal data were available, it was not possible to answer the question directly; however, one would suspect, with Stinchcombe[39] that there is a general tendency for organizational types, once founded, to show a great deal of structural stability over time relative to other types.

The question of structural stability, however, raises an interesting question about the six organizations with a rehabilitation orientation founded before 1946. Within the framework developed here, two alternative explanations, admittedly speculative, can be developed to explain their foundation before 1946. First, they may well have been early adopters or innovators, adopting an organizational form before it gained wide acceptance. The data for rehabilitation-oriented workshops approximate the first half of the S curve of adoption consistently reported in the innovation-diffusion literature. This explanation is consistent with the structural stability proposition, but would necessitate a partial revision of the basic hypothesis to include notions of organizational lead and lag. Environmental constraints change at varying rates, and the choice of a given time, for instance, 1946, as the point at which the constraints changed is somewhat artificial, being

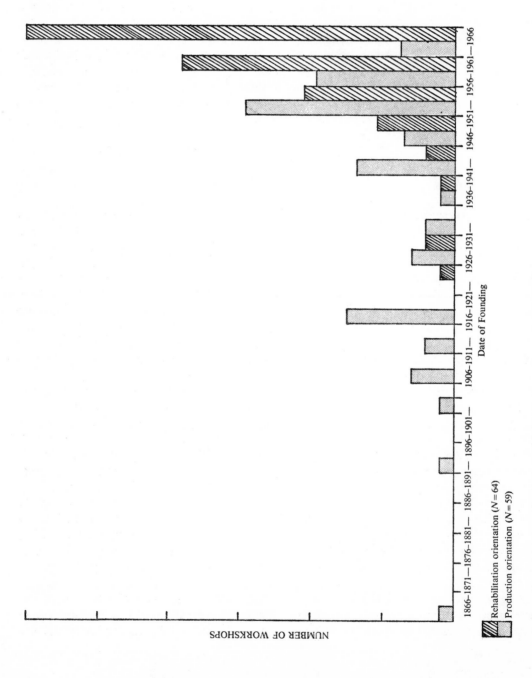

Exhibit 2-2.   Relationships between Orientation and Date of Founding

dictated by analytic necessity. It seems possible that organized responses to changes in environmental constraints occur at varying rates, with some organizations being founded in anticipation of these changes, and others probably most of them, being founded only after the changes have taken place and their impact is clearly understood.

A second explanation requires no revision of the basic hypothesis, but does suggest modification of the structural-stability proposition. It may be that the six organizations were able to develop structures capable of adapting to changes in external constraints; that is, they may have been production-oriented when founded, but later changed in response to changes in environmental factors. This explanation would require that the structural-stability proposition be revised to include the possibility of structural adaptability and some specification of the conditions likely to produce adaptive responses.

A different sort of question can be raised about organizational mortality. The data could be interpreted as illustrating higher mortality rates for organizations with a rehabilitation orientation. There are no data available to examine this question, but since production-oriented workshops rely heavily on income from goods-producing activities and are thus confronted by the necessity of coping with market conditions notoriously difficult for small businesses to deal with effectively, it is likely that the sample substantially underrepresents production-oriented organizations. Given the generally very high rates of failure in small businesses and the small size of workshops relative to other kinds of organizations, it is likely that many more production-oriented workshops were founded before 1946 than appeared in the sample. This interpretation would also help explain the fact that the data in the study show more production-oriented organizations being founded in the postwar era than

before. This most conservative estimate of the impact of organizational mortality on the findings, therefore, is that there is no reason to believe that the rates are any higher for rehabilitation-oriented than production-oriented organizations. A less conservative estimate is that the rates may be higher for production-oriented organizations. The effect on the interpretation of the data is thus conservative, and the theory is actually more strongly supported than the data suggest.

Influence of Environmental Constraints on Organizational Goals

Stinchcombe suggested that "an organization must have an elite structure of such a form and character that those people in the society who control resources essential to the organization's success will be satisfied that their interests are represented in the goal-setting apparatus of the enterprise"[40]. Because organizations cannot typically store all the resources they will need indefinitely, access to resources available only outside the organization becomes problematic at some point. At this point, the organization has to deal with another sort of environmental constraint.

In the case of sheltered workshops, the nature of these elite structures, here taken to mean the upper levels of management or what Thompson[41] has called the "dominant coalitions," might be expected to be related to when the organizations were founded, their orientation, and therefore to other structural characteristics. We would expect the elite structures of organizations with a rehabilitation orientation to reflect the newly emerging emphasis on the important of professional therapy and integration of the handicapped into the community. Given this emphasis and the greatly increased federal support available to sheltered workshops from the Barden-LaFollette Act and the amendments to the Vocational Rehabilitation Act, it might be expected that those individuals and agencies responsible for allocation of the newly available funds would be most likely to channel them to workshops with a rehabilitation orientation; that is, to those organizations whose elite structures, comprised basically of professionals, could provide reasonable assurance to decision makers that their own emphasis on and interest in rehabilitation would be respected. Therefore, one would expect that more workshops with a rehabilitation orientation would receive income from grants than workshops with a production orientation; furthermore, workshops with a rehabilitation orientation would be more likely to have income from grants than those with a production orientation regardless of when they were founded. Since a workshop was classified as production-oriented if it received 50 percent or more of its income from production activities, the hypothesized relationship might appear tautologous. However, the amount of income from grants received by all of the organizations in the sample accounted for only 5.4 percent of total income, and one would thus not expect differences to emerge as a consequence of the measure used to define orientation.

In order to test these hypotheses, several analyses were performed. First, the relationship between orientation and income from grants was examined. Of the workshops with rehabilitation orientation, 15 had no income from grants and 49 had some income from grants; of those with a production orientation, 39 had no income from grants and 20 had some[42]. These results indicate that workshops with a rehabilitation orientation are more likely to have income from grants than workshops with a production orientation. Type of orientation and income from grants are not independent.

To examine the combined effects of orientation and date of founding on income from grants, a three-way contingency analysis was performed. The results of this analysis are presented in Exhibit 2–3. From these results it appears that the effect of orientation is relatively independent of the effect of date of founding. Although analysis of variance would have answered this question more clearly, the fact that cases were unevenly distributed across cells and that orientation and date of founding were not independent suggested that use of this technique would be inappropriate. Instead, partial correlations were computed. With variable 1, date of founding; variable 2, orientation; and variable 3, income from grants, then $r_{13.2} = -0.14$ (n.s.) and $r_{23.1} = -0.44$ ($p < 0.01$). These results demonstrate quite strongly the relationship between orientation and income from grants. The multiple correlation for all three variables was 0.51.

Taken together, these findings illustrate one very important way in which the environment can impinge on the structure of an organization. The federal government's relatively recent involvement in active financial support of workshops helps to explain why the workshops founded after World War II have different structures from those founded before the war.

CONCLUSIONS AND IMPLICATIONS

This paper has presented the results of an attempt to investigate empirically and comparatively one aspect of the relationship between the orientations and environments of 123 rehabilitation organizations. Based on the findings of the study at least two conclusions can be drawn. First, empirical support was found for the interpretation of the Stinchcombe and Burns hypothesis about the relationship between social structure and organizational structure. This finding suggests the utility of a general theoretical perspective, which views organizational structure as the product of a set of interacting constraints, both internal and external, which are subject to varying degrees of direct control by organizational members. It would appear on the basis of the results here, that factors over which organizational members have very little direct control are very important in determining structural outcomes. Recent developments in the organization of health services in the United States suggest that external factors play an important role in other kinds of organizations as well. The increasingly large number of hospital mergers and the development of community-wide integrated health care delivery systems are two examples of attempts to reduce environmental uncertainty by increasing the amount of direct control by organizational members of certain segments of the environment.

The results also suggest that an alternative strategy for dealing with environmental uncertainty is development of structures and programs likely to increase organizational visibility to those individuals and agencies in society who control essential resources. This source of support, may, however, be somewhat ephemeral. If one accepts the Stinchcombe hypothesis about the tendency toward structural stability, then an organiza-

tion that relies heavily on this source of support may be unable to develop alternatives should priorities change and the support be withdrawn. Although the organizations included in this study did not rely exclusively, or even heavily, on federal funding, other organizations have and do, and their mortality rates tend to be high.

A second conclusion, admittedly speculative, concerns differential rates of organized responses to changes in values and technology. Not only are there lags between the development of values and technology and their institutionalization in forms that have the possibility of widespread impact, but there are also lags, which have been well documented, between the emergence of the new possibilities they represent and the development of organizational forms capable of ensuring that their impact will be widespread. One example might be the provision of adequate medical care to the poor. It would appear that social values supportive of the proposition that the poor should receive adequate medical care are highly developed and have been institutionalized legislatively; it would also appear that the technological capacity exists to make adequate care available. The problem now being confronted is how to organize resources in order to transform the possibility into a reality. Most of the research related to the

Exhibit 2-3. Relationship between Orientation, Date of Founding and Income from Grants

Orientation	Through WW II Income from grants		After WW II Income from grants	
	None	Some	None	Some
Rehabilitation	2	4	13	45
Production	19	8	20	12
	$n = 33$	$x^2 = 1.53$ n.s.	$n = 90$	$x^2 = 12.60$ $p < .001$
	$df = 1$	$r_\phi = .22$	$df = 1$	$r_\phi = .37$ $p < .01$

general problem underscored by this example is from the perspective of the individual rather than the organization. The results of the study presented here suggest the utility of an organizational perspective and point to the need to understand more clearly how external constraints affect the kinds of organizational forms that are likely to develop at any particular point in time.

While many questions remain, the findings underscore the importance of taking changes in the external constraints on organizations into account when attempting to explain differences in organizational structure. When the results are viewed in the light of results from other such efforts—for instance, Blau and Schoenherr[43]—it becomes clear that an open-systems approach to organizational analysis can be used to increase our understanding of organizations at the empirical level as well. It is also clear, however, that we have only just begun to do what is needed at this level.

Both Dill's early study of task environment and Kimberly's study conducted 17 years later discuss the constraining effects of environment on an organization. Dill described the effect of environment on performance; Kimberly described the effect of environment on structure. Kimberly suggests that after 17 years of research, organization theorists have only begun to do what is needed to acquire the empirical evidence needed for an understanding of organizations.

One might ask, "Now that I have this research, what can I do with the information?" The state of the art is such that definitive application of findings to daily decision making is not yet practical. The research should, however, raise your level of awareness to the point where you recognize the extent to which the external environment of an organization shapes an organization's structure and ability to perform. It should lead you to question whether you and your hospital personnel are well informed about changes in the external environment to which your department should be responding; such changes might occur in referral networks, health legislation, population distribution, medical technology, professional education, neighbor characteristics, licensure requirements, interagency alignments and mergers, treatment modalities, information retrieval, community salary norms, and so on.

Kimberly's study touched on strategies organizations can use to reduce institutional uncertainty. Responses of organizations to environmental constraints are discussed more extensively in Unit III.

8. The Sociocultural Environment

The selections offered in this volume thus far have described not only conflicts between internal organizational objectives and external sources of demand, but also conflicting demands that the external environment itself makes on the organization. We have seen how organizations respond to environmental changes by altering their structure and processes to fit the environmental contingencies and constraints. Research in the 1970s continues to struggle with the question of what particular combination of internal organization characteristics is most effective in permitting organizations to adapt to environmental contingencies and constraints[44]. As critical environmental factors are identified and studied, it should be possible for managers to plan strategies which enable the organization to respond to the environment effectively[45].

The reverse is much less feasible. As Richman and Farmer indicate, it is difficult, if not impossible, for most institutions to alter broad environmental conditions even though such macroenvironmental changes could lead to substantially better performance of organizations on a wider scale than is true when relying upon the cumulative effect of improvements initiated by individual organizations.

Generally speaking, the terms macroenvironment, broad environmental conditions, and generalized constraining influences refer to social, cultural, legal, political, and ecological forces of the environment which affect organization units.

Although broad environmental conditions are rarely altered by individual institutions, they are not to be ignored. Organizations, as boundary-maintaining, information-processing open systems, are in contact with the macroenvironment at many points. If administrators can understand those aspects of the macroenviron-

97

ment of significance to their particular institution, they can plan actions which fit the values, expectations, and requirements of the larger society. This is especially relevant for nonprofit institutions of the health care industry, most of which do not have access to the invisible hand of the market mechanisms which customarily keep profit-making enterprises informed of society's wishes and priorities at the marketplace. Many community sit-ins occurred at large city hospitals during the sixties because of the hospitals' failure to meet the expectations of the people served.

Organizational theorists have recognized a need to go beyond the study of the task environment to consideration of the impact of cultural and societal influences of an organization. Evans[46] suggests that the cultural and social structures of a society probably account for an appreciable amount of the variance of any major dependent variable, such as organization performance. He adds that the organizations capable of modifying the environment engage in essentially political actions; that is, they use resources and sanctions against actual and potential opponents to alter events.

The selections which follow describe how social, cultural, and political forces influence organization structure, transactions, and strategies. No attempt is made to enter into a theoretical discussion concerning the concepts of culture, social structure, or politics; rather, the selections focus, in a general way, on how organizations deal with the values, norms, institutional expectations, and legal mechanisms of their external environments. By now, the reader should be alert to the value of identifying what resources flow across the institution's boundaries and the characteristics of these transactions. Knowledge of the cultural, political, and social context within which institutional transactions occur will further the administrator's understanding of the conditions under which organizations take the initiative to modify their social arrangements and thereby further their objectives.

EFFECT OF POPULATION AND HEALTH CARE ENVIRONMENT ON HOSPITAL UTILIZATION

By Daniel M. Harris

Reproduced with permission from *Health Services Research*, Volume 10, Number 3, Fall 1975, pp. 229-243. Copyright 1975 by the Hospital Research and Educational Trust.

The next selection differs from the preceding ones in that it does not directly address the theoretical base of modern organizational research. Rather, it is a sociological study which uses a direct causal model to analyze a difficult sociological question. It studies how society's demographic and socioeconomic characteristics determine usage patterns of hospitals, the input of the organization.

Hospitalization is a social as well as a medical phenomenon. The way people use hospitals depends largely on the type of people they are, the way medical care is delivered to them, and the type of medical resources available to them[47].

A population's age and sex distribution, marital status, fertility experience, density, and occupational structure are all related to its morbidity level. Its income, educational, and racial and economic composition in large part determines the manner in which it responds to this morbidity experience, as well as its ability to respond. Thus a population's demographic and socioeconomic characteristics objectively determine its health care requirements.

Health care requirements, however, can be met in different ways or not at all. How these requirements are met depends in large part on the health care environment of a given population. In other words, the health care environment intervenes between population and hospitalization. A causal sequence is implied here: different combinations of population characteristics lead to different types of health care environments, which in turn lead to different usage patterns of hospitals and other medical facilities.

The health care environment can be divided into two components. One is the health care delivery component, consisting of the location, control, and organization of medical facilities, the types of services offered by different facilities and providers, and the like. The other component, health care resources, refers to the number of physicians and other health care personnel, the number of hospital beds, the number and kinds of alternatives to inpatient hospital care, and the like. If health care delivery characteristics determine the level of health care resources that can be supported, then the causal model becomes: population characteristics leading to health care delivery charcterisitcs leading to level of health care resources leading to hospital utilization patterns. This model, then, should allow one to predict and account for a population's rate of hospital use.

This article investigates the accuracy of the proposed causal model in predicting utilization of short-term general hospitals. The unit of analysis is the resident population of a county. Data to test the model were collected from 56 New York State counties and were analyzed through multiple correlation and path analytic techniques. A review of the relevant literature suggested which variables to include as independent variables in the model. In addition, factor analysis was used to construct multivariable indicators from combinations of the large number of highly interrelated population characteristics.

RESEARCH DESIGN

Population Variables

Past research has indicted that many population characteristics are related to hospital utilization. Wirick, for example, found that age, income, sex, and education help determine a **person's** chances to be hospitalized[48]. Anderson, on the other hand, cited not only these variables but also ethnic composition, unemployment, migration, urbanization, and percent of labor force in agriculture as determinants of hospital utilization for the resident population of New Mexico **counties**[49]. Similar findings using these variables, and either the individual or aggregate population as the unit of analysis, have been reported by others[50].

Studies in which the aggregate population is used as the unit of analysis generally do not take into account how the various population characteristics actually combine and how these combinations affect the overall utilization of hospitals; for example, married females in the childbearing years have one of the highest hospital utilization rates due to childbirth and complications related to pregnancy, but populations with high proportions of married females between the ages of 15 and 49 also tend to have high proportions of children, the age group that experiences the least hospitalization. The combination of these two population groups and their characteristic utilization patterns results in an aggregate utilization rate quite different from the utilization rate of either group considered by itself.

To deal with such combinations of factors, as well as to reduce the large number of previously cited socioeconomic and demographic characteristics, a factor analysis was performed. This technique helps identify underlying common dimensions in a set of independent variables. The factors that are thus extracted can be interpreted on the basis of the loadings of variables on them. The loading, or contribution of a variable to a factor, is the zero-order correlation coefficient between a variable and a factor. As with all correlation coefficients, the square of the coefficient is the proportion of covariance or overlap. A loading of 0.316 or higher—that is, at least 10 percent covariance—is taken as being substantively significant and requiring interpretation.

Exhibit 2-4 presents the results of a factor analysis of 21 demographic and socioeconomic variables for 56 New York State counties. These counties are all the counties in New York excluding the five that comprise New York City and one that had no hospital as of 1970. The data come from statistics published by New York State[51] or the U.S. Bureau of the Census, and all refer to either 1969 or 1970. The table contains the factor loadings for each of the 21 population characteristics and the communality estimates (h^2) for each characteristic. The communality figures estimate the proportion of each population variable's variance accounted for by the factor analysis. Most of

Exhibit 2-4. Factor Analysis of Population Characteristics for 56 New York State Counties (Definitions from 1970 Census)

Variable	Factor				h^2
	1	2	3	4	
1. Median family income	**0.927**	−0.243	0.034	0.071	0.924
2. Personal income per capita	**0.876**	0.096	−0.210	0.099	0.830
3. Population per square mile	**0.867**	0.014	−0.179	−0.077	0.789
4. Size of population	**0.827**	−0.050	−0.089	0.160	0.720
5. Percent of families below poverty level	**−0.774**	0.205	−0.059	−0.129	0.661
6. Percent of work force in white-collar jobs	**0.771**	−0.308	−0.258	0.158	0.780
7. Percent of population living in urban areas	**0.738**	−0.135	−0.107	**0.483**	0.807
8. Percent nonwhite	**−0.674**	−0.018	−0.092	0.149	0.486
9. County located in SMSA	**0.560**	−0.230	0.241	**0.514**	0.689
10. Percent of population over 25 who have completed 13 or more years of school	**0.551**	**−0.548**	**−0.391**	−0.127	0.773
11. Crude death rate	**−0.497**	**0.782**	−0.194	0.083	0.903
12. Percent of population age 65 and over	**−0.429**	**0.811**	−0.287	0.083	0.930
13. Median age	0.252	**0.894**	−0.174	0.094	0.902
14. Percent reproductive age females (15–49)	0.251	**−0.797**	**0.438**	0.121	0.904
15. Percent of population under age 5	−0.121	**−0.441**	**0.846**	0.044	0.926
16. Crude birth rate	**−0.477**	**−0.407**	**0.576**	0.165	0.752
17. Fertility ratio†	−0.314	0.018	**0.905**	−0.099	0.927
18. Percent females (14 and older) married	0.213	0.256	**0.795**	−0.346	0.862
19. County has central city	0.249	−0.109	−0.056	**0.681**	0.540
20. Sex ratio	−0.176	−0.170	0.302	**−0.617**	0.531
21. Infant mortality rate	−0.059	**0.335**	0.038	**0.494**	0.361

*Variables in each factor are indicated by boldface type
† Number of children age 5 and under per female between 15 and 54 years of age.

these estimates were very high, with only two, those for percent nonwhite and infant mortality rate, below 0.50.

The four-factor orthogonal rotation solution shown in Exhibit 2-4 accounted for 76 percent of the total variance in all of the population variables. Each factor accounted for at least 9 percent of the total variance (additional factors accounted for less than 5 percent each). The first factor was defined primarily by the large contributions of the first 10 items. Since these items all refer to metropolitan residence or high socio-economic status, this factor was named "metropolitan/middle class." Items 11, 12, and 16 also loaded substantively on this factor. These loadings were negative, fertility and mortality rates both being lower in middle class than in working class populations.

Items 11–14 had at least 60 percent common variance with factor 2. These items all relate to age and chronic illness. Education, percent of population under age 5, and crude birth rate (items 10, 15, and 16) were all negatively associated with this factor and shared between 17 and 30 percent common variance with it. These items, along with item 21 (infant mortality rate), which had over 10 percent of its variance accounted for by this factor, can all be interpreted in terms of advancing age and increased morbidity. Factor 2 was named "age/illness."

Factor 3, which had high positive loadings for items 14–18, is a "fertility/family" factor. Fertility is somewhat negatively related to education; item 10 had a loading of –0.391 on this factor.

The final factor was identified primarily by items 7, 9, 19, 20, and 21. Counties that scored high on this factor tend to be urban, to contain an SMSA central city, and to have a high proportion of women, many of whom are married (negative loading of item 18). This result supported a city/inner suburb in-terpretation of this factor since these areas attract young, single women for employment and marriage opportunities and hold onto older widows. Factor 4 was thus named "city/inner suburb."

Factor scores, or scores on the dimensions identified by the factors, for each of the 56 counties on each of the four factors were computed, adjusted for negative values, and used in the subsequent analysis.

Health Care Delivery System Variables

The next step in the causal model is the county's health care delivery system. For the purpose of predicting hospitalization rates, two variables were chosen to describe this system: mean size of short-term general hospitals and concentration of short-term general hospital beds.

The size of a hospital has been shown in the past to be associated with level of utilization[52], because "size is assumed to be related to a number of other factors such as scope of services, specialization of medical staff, and presence of educational and research programs that affect the pattern of use of hospital services"[53]. The size variable was measured with data from the American Hospital Association's listing of hospitals for 1970[54] and was operationalized as the arithmetic mean of the number of beds in all short-term general hospitals in a given county.

To test the relationship between hospital size and utilization for the hospitals included in the present study, data from a previous study of New York State hospitals were used[55]. For a sample of New York State not-for-profit short-term general hospitals, the coefficient of correlation between size (measured as number of beds) and number of medical facilities was 0.78 and the coefficient of correlation between size and number of educational, research, and medical programs was 0.71. Further, a factor analysis of a sample of New York State county health work forces, performed for a separate purpose, produced a professionalization factor that had a coefficient of correlation with mean hospital size of 0.49.

Distance to the site of the hospital also has been cited in the literature as affecting hospital utilization[56]. This is a difficult variable to measure for an aggregate population; however, it was felt that if all or most of the short-term general hospital beds were concentrated in one hospital, then the distance factor would be maximized. Consequently a bed concentration variable was derived and operationalized as the number of beds in a county's largest short-term general hospital expressed as a percentage of the total number of beds in that county's short-term general hospitals. The higher the percent, the more concentrated the bed supply. As a partial check on the validity of this variable, it was correlated with bed density (the total number of beds in a county's short-term general hospitals per square mile). The correlation coefficient was –0.48, meaning the higher the bed concentration, the fewer beds per square mile. Thus the higher the bed concentration, the greater the average distance from all points in the county to that county's hospital beds.

Health Care Resource Variables

The causal model next cites health care resources as affecting hospital utilization. Three variables were used to measure resources: number of physicians in a county per 100,000 county residents, total number of beds in a county's short-term general hospitals per 1,000 county residents, and the availability of at least one organized short-term general hospital outpatient department in the county.

The first two variables, both of which control for population size, have been cited in the past as influencing hospital utilization[57]. Rosenthal[58], however, while admitting the association between resources (supply) and utilization (demand), believes the relationship to be spurious. A high level of supply, he writes, "arises as a response to a number of pressures, one of which could be the demand for facilities. If demand for facilities did exist, the factors that affect demand would also generate pressure to affect supply."

Does demand lead to supply, as Rosenthal states, or does supply lead to demand, as the causal model under consideration indicates? The causal priority of these factors can be tested using the technique of cross-lagged partial correlations as described by Pelz and Andrews[59]. Using data for the 56 New York counties under study, I calculated the partial correlation coefficient between a county's 1960 supply (hospital beds per 1,000 residents) and its 1970 demand (patient-days per 1,000 residents), controlling for the 1960 demand, and the value was 0.352. Between 1960 demand and 1970 supply, however, the coefficient of partial correlation (with 1960 supply controlled for) was only 0.084. This can be interpreted as meaning that supply has a greater causal priority than demand.

Furthermore, a recent study[60] has shown that high demand in 1960 for a county's short-term general hospitals was unrelated either to changes in hospital bed supply or to changes in hospital utilization between 1960 and 1970; whereas changes in a county's short-term general hospital bed supply led to changes in its admission rate and length of stay, which affected average daily census rates in these hospitals. From this evidence, then, supply of medical resources should precede hospital utilization in a causal model linking the two.

The third variable, presence of an outpatient department, has also been cited as affecting hospital utilization. Ro, for example, states: "Substitution of outpatient care . . . for inpatient care takes place when those substitutes are available . . . . Patients treated at hospitals which have outpatient clinics were hospitalized for shorter periods . . . than those treated as hospitals without outpatient clinic facilities."[61] This variable was measured as a dummy variable having the value of 1 if the county had at least one short-term general hospital with an organized outpatient clinic in 1970 and value of 0 if no short-term general hospital in the county had such a clinic (data for this variable came from the American Hospital Association[62]).

Measures of Utilization

According to the causal model, all of the preceding variables affect the utilization rate of a county's general hospitals. Utilization was measured by three variables: total 1970 admissions to a county's short-term general hospitals per 1,000 residents of that county in 1970, the total average daily census in a county's short-term general hospitals in 1970 times 365 days divided by the total number of 1970 admissions to these hospitals (this measures average patient-days per admission and is an estimate of average length of stay), and the total 1970 average daily census in a county's short-term general hospitals per 1,000 residents of that county in 1970. These variables, or variants of them, are among the most common measures of hospital utilization in the literature[63]. The first two variables are components of the third one and precede it in the causal analysis.

Since not all of a short-term hospital's inpatients reside in the county in which the hospital is located, utlization measures in which the number of residents in the county is used as the denominator tend to inflate estimates of utilization by county residents. Fortunately, the size of this measurement error can be estimated from a previous study[64], in which administrators of short-term general hospitals in New York State, excluding those in New York City, were asked to estimate what percentage of their inpatients came from the county in which the hospital was located. For the 152 administrators who responded, the average estimate was 86 percent. Thus the average error for the present study should be small.

PATH ANALYSIS

The causal model was then tested by means of path analysis. This technique is an extension of multiple linear regression that allows the researcher to causally order the effect of independent variables believed to affect a dependent one and then to assess the accuracy and utility of the causal model so constructed[65]. Path analysis allows one to measure the total effect of an independent variable—its direct effect on a dependent variable plus its indirect effects through changes it causes in other intervening independent variables. In terms of the causal model under investigation here, the total effect of the four population factors on hospitalization rates equals the direct effect (not involving other variables) of these factors on hospitalization plus the indirect effects due to the effect of each factor on the two health care delivery and three health care resource variables and the effect of these five variables in turn on the three utilization measures. The same argument can be constructed for the total effect of the two health care delivery variables on utilization: a direct effect and indirect effects through the three health care resources variables.

Path analysis proceeds by solving a series of regression equations in succession, making each variable at each causally subsequent step the dependent variable in turn. The effects of the independent variables at causal step 1 on those at step 2 are first assessed through ordinary multiple regression, the standardized partial regression coefficient (path coefficient) measuring the effect of each causally prior variable on each causally subsequent one. Next, the effects of all the variables at steps 1 and 2 on those at step 3 are assessed in the same manner, and so on until all causal steps are assessed.

The total association between a causally prior variable and a causally subsequent one is measured by their zero-order correlation coefficient. This total association equals the total effect (direct effect plus any indirect effects) plus 1) any spurious effects due to causally prior variables or 2) joint associations due to collinearity with variables at the same causal step.

The direct effect is measured by the path coefficient directly connecting the two variables. Each of the indirect effects is estimated by multiplying together the path coefficients indirectly connecting the causally prior variable to the subsequent one through one or more intervening variables.

The overall utility of the causal model is assessed through its predictive power. This is estimated by the square of the multiple correlation coefficients associated with each of the successive regression equations. These squared coefficients measure the proportion of variance in each of the successive dependent variables that is accounted for by the causally prior variables included in the regression equation. The proportion of vari-

ance unaccounted for, then, equals 1 minus the squared correlation coefficient. Unaccounted-for variance is assumed to be due to the influence of variables left out of the causal model and to measurement error.

The path analysis performed here followed the convention suggested by Duncan[66] to eliminate all path coefficients that are not statistically significant. The level of significance chosen for this study was $p < 0.10$.

RESULTS

Exhibit 2-5 presents the zero-order correlation matrix for the 12 variables included in the path analysis test of the causal model. The matrix is arranged so that reading down the columns shows the correlations between a given variable and either variables at the same stage in the causal model (marked with an asterisk in the table) or variables that precede it in the model. The four socioeconomic and demographic factors are at the bottom of the matrix, the two health care delivery characteristics are above them, the three health care resource variables are next, and the three utilization variables are at the top.

None of the 11 correlations between variables at the same stage in the model exceeded 0.28. This means that joint associations added very little to the total associations. The four population factors were uncorrelated, as expected since an orthogonal rotation was used. Mean hospital size and bed concentration had a correlation of −0.19. Outpatient department and bed supply were largely unrelated, but the correlations between doctors per 100,000 residents and bed supply and outpatient department were 0.28 and 0.26, respectively. Admissions and length of stay were correlated at only 0.13.

Exhibit 2-6 shows the results of the path

Exhibit 2-5. Correlation Matrix for Variables Included in Path Analytic Test of the Causal Mode ($N = 56$ Counties)

	1	2	3	4	5	6	7	8	9	10	11	12
1 Census/ 1000 population	. . .											
2 Admissions/ 1000 population	.93	. . .										
3 Length of stay	.46	.13*	. . .									
4 Physicians/ 100,000 population	.34	.37	.05	. . .								
5 Beds/1000 population	.96	.89	.44	.28*	. . .							
6 OP department available	.15	.03	.32	.26*	.10*	. . .						
7 Mean hospital size	.44	.44	.13	.63	.34	.21	. . .					
8 Bed concentration	-.24	-.20	-.15	-.41	-.25	-.48	-.19*	. . .				
9 Metro/middle class (factor 1)	-.13	-.13	-.05	.60	-.18	.40	.35	-.53	. . .			
10 Age/illness (factor 2)	.27	.21	.21	-.05	.29	-.02	-.23	-.10	-.00*	. . .		
11 Fertility/family (factor 3)	-.24	-.16	-.27	-.37	-.23	.00	-.22	.04	-.02*	.01*	. . .	
12 City/inner suburb (factor 4)	.16	.16	.08	.25	.10	.32	.40	-.26	.01*	-.00*	-.00*	. . .

* Correlation between variables at the same stage in the causal model.

analysis. The numbers on the causal arrows are the path coefficients. The arrows without origins represent the influence of unmeasured variables and measurement error. The numbers on these arrows equal the square root of the variance unaccounted for (1 minus the multiple correlation coefficient squared).

Determinants of Utilization

According to the results of this analysis a population's general hospital bed supply is the strongest determinant of its rate of hospital utilization. Bed supply has the largest total association with all three indicators of utilization, the largest direct effect

on admissions and length of stay, and the largest indirect effect on hospital census. Five of the six variables that are causally prior to bed supply affect utilization indirectly through this variable. Only factor 3, fertility/ family, bypasses bed supply and directly affects admissions and length of stay.

The other health care resource variables are also important. The results show that the higher the physician/population ratio, the higher the admissions per 1,000 residents, and through this the higher the hospital census per 1,000 residents. However, a large supply of doctors decreases length of stay, and through this causal mechanism decreases census per 1,000 residents. A com-

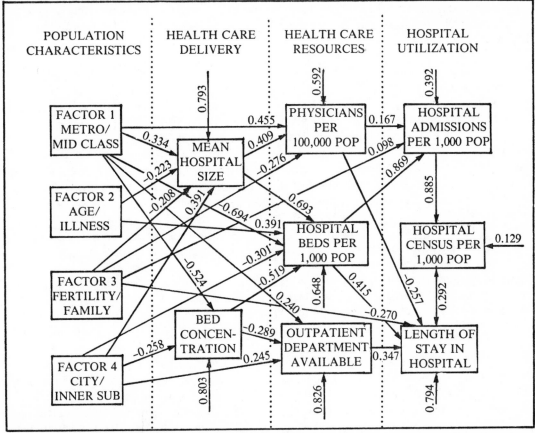

Exhibit 2-6. Path diagram for test of causal model.

parison of the two indirect effects, though, reveals that the net indirect influence is +0.06. Thus the presence of high physician/ population ratios increases admissions while decreasing lengths of stay (indicating the admissions of cases of less severity) and leads to a net increase of patient-days of hospital use.

The results for outpatient departments disagree with the results reported by Ro[67] as well as with what was originally expected. However, Ro's unit of analysis was the individual rather than the aggregate population, so his results are not entirely comparable with those reported here. It appears that a hospital outpatient clinic may be used

to treat cases of low severity that, in the absence of the clinic, might otherwise have been admitted to the hospital. This frees hospital beds to be used for cases of greater severity that require longer lengths of stay (path coefficient 0.347). The next effect is to slightly increase hospital census per 1,000 residents.

Determinants of Health Care Resources

Counties with a large mean hospital size (that is, with many services and facilities and a highly specialized health work force) attract a large supply of physicians per 100,000 residents and are able to maintain a large supply of beds per 1,000 residents. As shown, the physicians tend to make use of the beds available to them; thus hospital size affects hospital utilization through the health care resources that size can support.

Counties with concentrated hospital beds maintain fewer beds per 1,000 residents (path coefficient -0.519) and have a smaller tendency to possess outpatient departments (path coefficient -0.289) than counties with more dispersed beds. The net indirect effect is to decrease hospitalization rates, but it is not clear whether this is due to average distance to the hospital or to the lack of abundant health care resources characteristic of counties with concentrated bed supplies. The small negative correlation between bed concentration and mean hospital size (-0.19), however, is indicative of the latter rather than the former explanation.

The population factors have interesting patterns of direct and indirect effects on hospitalization rates. First, only factor 3 (fertility/family) has direct effects on the utilization measures. This factor has a very small positive direct effect on admissions (0.098), probably the result of the high admission rate of women in the childbearing years being counteracted by the low rate of their offspring, thus resulting in a very small path coefficient. It also has a negative direct effect on length of stay (-0.270), probably due to the short stays of women of childbearing age. The net indirect effect on hospital census per 1,000 residents through these two intervening utilization measures is -0.007. In other words, the two opposite direct effects essentially cancel each other out.

The only remaining effects of factor 3 on utilization come from the indirect effects through mean hospital size and physicians per 100,000 residents. The direct effects on

both of these intervening variables are negative; thus the overall net indirect effects on utilization are negative, though small (-0.16).

The overall influence on utilization of factor 1 (metropolitan status/middle class characteristics) is somewhat similar to the influence of factor 4 (central city/inner suburb characteristics). Both factors have positive direct effects on mean hospital size (thus positive net indirect effects on utilization through this variable), negative direct effects on bed concentration (thus positive net effects on utilization through this variable), and positive direct effects on the existence of a general hospital outpatient department (and thus positive net indirect effects on utilization through this variable as well). In addition, factor 1 also has positive net indirect effects on utilization through its positive direct effect on physicians per 100,000 residents.

However, factors 1 and 2 both exert a negative direct effect on hospital beds per 1,000 residents. In the case of factor 1, this effect is quite strong (path coefficient -0.694). The net indirect effects through bed supply for both of these factors, then, is negative. For factor 1 this negative effect is so strong that it slightly outweighs the positive indirect effects, leaving factor 1 with a negative total net indirect effect on utilization (-0.09). Factor 4 ends up with a positive overall net indirect effect (+0.14), but it is greatly reduced from what it would have been if the direct effect on bed supply were not negative.

The remaining population factor, factor 2 (age/illness), has no direct or indirect effects on either bed concentration or outpatient department availability. It affects utilization indirectly only through mean hospital size and bed supply. The negative net indirect effect through size is more than offset by the positive net indirect effect through beds, yielding a positive net indirect effect (+0.21).

DISCUSSION

The results generally support the causal model. The population factors affect health care delivery characteristics and account for 34 percent of the variance in bed concentration and 37 percent of the variance in mean hospital size. The combination of population and health care delivery characteristics accounts for 64 percent of the variance in the supply of physicians, 57 percent in the supply of short-term general hospital beds, and 31 percent in the availability of a general hospital outpatient department. All of these variables, in turn, account for 82 percent of the variance in admission rates and 35 percent of the variance in length of stay. In addition, the combination of the effects of the two principal components of census, admissions and length of stay, accounts for over 98 percent of the variance in average daily census (patient-days) per 1,000 residents.

The most striking result of this analysis is the central importance of a county's bed supply in determining the utilization of its short-term hospitals by its resident population. Although this relationship has been noted in the past, this analysis grounds it in the context of a causal model that links population, health care delivery, and health care resource characteristics with hospital utilization.

Locating this relationship in a causal model allows a fuller understanding of the mechanics of the relationship. Bed supply and physician supply are somewhat postively associated (correlation of 0.28), and mean hospital size is causally related to both of these resource variables. Increases in hospital size, then, lead to increases in the level of supply of both physicians and beds, which in turn intensify the increase in hospital utilization. Furthermore, increasing bed supply increases both admission rate and length of stay, although it is through admissions that bed supply has its strongest effect on census.

Looking at the model as a whole reveals an interesting pattern of direct and indirect effects involving metropolitan and middle class characteristics, hospital bed supply, and hospital utilization. The net effect of metropolitan and middle class characteristics is a low utilization rate **because of the relative low supply of beds per 1,000** in counties scoring high on this factor. Increasing the bed supply even slightly in these counties, therefore, can be expected to lead to large increases in utilization, especially if these beds are added to the already existing smaller hospitals (thereby both further increasing the already high mean hospital size and further decreasing the already low bed concentration). Also, the addition of hospital beds in these counties can be expected to lead to greater increases in hospitalization than addition of an equivalent number of beds in more rural and less middle class counties.

Another striking result is the apparent absence of any strong direct causal connection between factor 2 and hospital utilization. The strongest indirect effect links age/illness to a large bed supply, which in turn leads to high utilization. If age/illness is seen as a measure of the demand for hospital services, then high demand is associated with both high bed supply and high utilization and the supply-utilization relationship is spurious, as Rosenthal has suggested[68]. However, the older populations characteristics of counties scoring high on factor 2 may be partly the result of an outmigration of young adults and any young children that they may have. If this is the case, then the relatively high bed/population ratios in these counties may be attributable in part to a decreasing population base rather than a high absolute number of beds.

Of the 56 counties studied, four decreased in population between 1960 and 1970. The mean score on factor 2 for these counties was 0.961. Since a factor score is a normalized score with a mean of 0 and a standard deviation of 1, this means that the four counties that lost population had a mean score on factor 2 almost a full standard deviation higher than the mean score of all 56 counties. When the score on factor 2 is correlated with the ratio of 1970 population to 1960 population, the resulting correlation coefficient is -0.40. Thus counties scoring high on factor 2 tended to have either decreased in population between 1960 and 1970 or increased less than counties that scored low on this factor.

Furthermore, the reasoning behind the spurious interpretation would lead one to expect a significant positive direct relationship between factor 2 and physicians per 100,000 residents. In fact, one does not exist. Also, a high demand should be associated with larger hospitals, which have the most medical facilities and services. Factor 2 is in fact associated with smaller hospitals. On the basis of these results, then, the spurious interpretation is not confirmed.

Implications and Conclusions

The most immediate practical implication of these results is that it is possible to alter hospital utilization rates by altering supplies of physicians and hospital beds. This is not the first time that this implication has been drawn—it was first suggested by Shain and Roemer in 1959[69] and again by Roemer in

1961[70]. However, as recently as 1969, Klarman could still write that "for the United States, Roemer's proposition does not command general assent even now"[71]. Even more recently, Andersen and Newman stated that, based on a general review of the literature and several specific empirical studies, community resources have low relative importance in determining hospital use[72].

The results of the present study, coupled with those of a companion study, using longitudinal data[73], should help end the controversy surrounding Roemer's thesis, at least in urbanized areas. It should now be clear that supply can create its own demand rather than that demand leads to congruent levels of supply. To supply additional beds to areas with current high demand (utilization) in an attempt to "satisfy" the "unmet need" in these areas is thus seen as a futile exercise. Additional beds will always lead to additional use, and health care expenditures will continue to rise. To base hospital construction priorities primarily on past use and current demand is hopeless since areas will never have "enough" beds.

Excessive hospitalization and its high costs can be reduced by controlling the number of beds available to a local population and its physicians, as well as by controlling the supply of physicians. Money that would have been spent on hospital stays can then be used to help finance and develop more rational health care delivery systems that make more extensive use of ambulatory care and other substitutes for inpatient care[74].

This research should also help make clear to health care personnel that hospital use is a social process embedded in a web of social causation. Morbidity levels and demographic characteristics have less effect on a population's rate of hospitalization than the way that population is organized to deliver health care services and the resultant supply of health care resources. More empirical social research needs to be done to better understand the working of these local health care environments.

Harris's research explains how different population characteristics lead to different types of health care environments, which in turn, lead to different utilization patterns of hospitals and other medical facilities. He factored the population characteristics into four sets of variables: metropolitan/middle class; age/illness; fertility/family; and city/ inner urban. Harris substantiates Roemer's long-standing thesis that in urban areas at least, the supply of hospital beds creates its own supply of patients. Although the purpose of including the study in this anthology is to offer an empirical investigation of the impact of social influences on organization (i.e., their utilization), the research also indicates the rationale behind the federal government's attempt to contain costs by limiting hospital beds and encouraging alternative delivery systems. Federal political actions such Public Law 93-641 affect hospital's power to control their flow of resources.

Harris offers empirical evidence that society's use of hospitals is based on "a web of social causation." In the next section, the selections will illustrate ways in which political environments affect organization performance.

9. The Political Environment

Organizational literature states that political conditions in the external environment can contribute to an organization's uncertainty and problems, as well as open up new opportunities for it[75]. In the case of interorganizational research, discussion of political environments centers around interorganizational networks and how member organizations challenge organizational domains in the acquisition or defense of an adequate supply of resources[76]. Aldrich[77] identifies the domain of the organization as a subset of the task environment. Domain refers to the range of activities claimed by the organization for itself as its particular arena of operation; i.e., the claim may be resisted by other elements in the task environment, and conflict may arise over the question of domain legitimacy. If one accepts the broad definition of politics as being concerned with the control and management of people living together in society, it is apparent that political environments directly influence organizational performance.

In the literature, the political aspect of the environment is frequently linked with other environmental dimensions; thus, we have Richman and Farmer's[78] "political-legal" environmental constraints, Schensul and Bymel's[79] "socio-political" community power, and Benson's[80] "political economy."

In describing an interorganizational network as a political economy, Benson states that interactions and sentiments of organizations are dependent upon their respective market positions and power to affect the flow of resources. The interorganizational network is itself linked to a larger environment consisting of authorities, legislative bodies, bureaus, and publics. The flow of resources into the network depends upon developments in this larger environment.

Two selections were chosen to illustrate environmental effects of a political nature. The first is a study describing the changes brought about in a West Side Chicago community mental health center as a direct result of sociopolitical power exerted by community residents. The center survived because it was able to be fitted to the sociocultural components of its external environment. Note that confrontation tactics were not used when the clinic failed to respond to community needs; rather, the Mexican residents simply avoided contact with the clinic. Only after the clinical staff worked directly with the activist groups on a broad variety of issues was the link forged which allowed the clinic to become a valuable resource to the community.

THE ROLE OF APPLIED RESEARCH IN THE DEVELOPMENT OF HEALTH SERVICES IN A CHICANO COMMUNITY IN CHICAGO

By Stephen L. Schensul and Mary Bakszysz Bymel

Reprinted with permission from *Topias & Utopias in Health*, edited by S. Ingman and E. Thomas. Published by Mouton Publishers, 1975.

In 1963 Congress enacted legislation to establish "community" mental health programs in urban and rural sites throughout the country. One such program was established in September of 1967 in Chicago's West Side Medical Complex. Its objective was to provide mental health services to the adjacent black community and to a large area immediately to the south in which Mexicans were rapidly replacing long-term residents of middle-European origin. To accomplish the task, four outpatient "storefront" clinics were set up in each of the main communities of the catchment area and were linked to inpatient and specialized services available at the medical center. These outpost clinics were to extend psychiatric services to people who until then had had little access to such care. From a community base, these centers were to mobilize community forces to help in the care and rehabilitation of patients and to effect positive changes in community structure so that mental illness could be prevented as well as treated.

The first of these outpost clinics was established in El Barrio (a pseudonym), a predominantly Mexican community located on Chicago's Near West Side. The El Barrio Mental Health Clinic began its operations at a time when few service institutions in Chicago had come to terms with the fact that there was a large population of recently immigrated Mexicans requiring new programs, resources, and services.

For the residents of El Barrio, this situation was reflected particularly in health care. A few local physicians and an overloaded county hospital were the main health resources for El Barrio residents. Differences in language, attitudes, and health practices among Mexicans were not understood by the medical establishment, creating additional barriers to effective uses of even these limited resources. The Community Mental Health Program had a difficult task in trying to establish a mental health service in a community with limited services and no experience with Anglo-American mental health concepts.

In the beginning of 1969, the director of the program hired the authors to establish an anthropologically oriented Community Research Unit for the purpose of providing information concerning the Mexican, middle-European, and black populations in the area. This research was seen as providing a base for planning new clinical programs designed specifically for Mexican residents and constructing preventive programs to integrate mental health services into community development. Our research unit was to collect information on the "natives" of the area so that plans, policies, and therapeutic methods could be developed by the program staff which would meet the special cultural and community needs of the area. This position closely parallels the traditional role of the applied anthropologist—that of a provider of information to dominant policy-making and power sectors on behalf of economically and politically margainal groups in a society.

Over the past five years a series of events, both in the community and in the program, changed this role drastically and created a situation in which our research unit collaborated directly with community groups in El Barrio in formulating independent health projects. In describing this collaborative process we will examine the events and actions that led to the establishment of community health programs and the strategies used by our community research team to facilitate these developments.

COMMUNITY LIFE AND HEALTH ISSUES IN EL BARRIO

The El Barrio community is Chicago's "port of entry" for Mexican immigrants and its major residential enclave for Mexican-Americans. A great majority of the Mexican population has arrived in Chicago within the last ten years directly from Mexico. Approximately 20 percent are from Texas and only 5 percent of individuals in a recent survey are Chicago-born[82]. These different origins in various segments of the population—"Tejanos," Mexican nationals, and Chicago-born Chicanos—produce differences in attitudes, life experience, and behavior.

The influx of Mexicans into this area continues a long history of a succession of immigrant groups. Prior to the 1880s Irish and German immigrants and native-born Ameri-

cans came to work in the small industries that were located in the area. Poles, Czechs, Slavs, and other middle-European groups began to enter the area in the mid-1880s and by 1900 were the predominant ethnic groups. For the succeeding fifty years the community maintained a strong middle-European ethnic character—one which has left a visible mark even now on the community.

With the construction of a university campus north of El Barrio, many displaced Mexicans began to move south into the houses vacated by the outgoing middle-Europeans. This movement, combined with a growing influx of people from Mexico and the American Southwest, resulted in a rapid increase in the Mexican component of the population. By the end of the 1960s, the Mexican-American sector had increased from 30 percent to almost 70 percent of El Barrio's population. Most of the migrants of this period came directly from Mexico from cities such as Monterrey, San Luis Potosi, Guadalajara, Michoacan, and other urban areas in western and northern Mexico.

The El Barrio community has a total population of 44,660 people, including 35,750 Mexicans, 2,211 Puerto Ricans, 5,631 middle-Europeans and 1,068 blacks. Low rents (averaging $88 a month), a close proximity to places of potential employment, rapidly decreasing numbers of middle-European residents, and the availability of Mexican goods and services make El Barrio a highly appropriate area for settlement by in-migrating Mexicans.

The overwhelmingly Mexican character of El Barrio permits the recent migrant to interact in Spanish in most contexts. Ethnographic data indicate that in Chicago a number of job situations exist which do not require English language abilities. Contacts through relatives and friends can lead to relatively satisfactory incomes through employment in Spanish-speaking work crews.

Spanish is the major language in most of the commercial establishments in the area. Availability of Spanish-language newspapers, magazines, music, household items, and food such as large quantities of **carnitas, chicharon,** and **pan dulce** convey a strong feeling of old Mexico.

El Barrio is viewed by many in Chicago as a typical inner-city "ghetto." In a recent article on economic and social status in Chicago, one of the city's newspapers rated El Barrio as eighty-fourth in a ranking of eighty-five Chicago communities. This rank was based on such indicators as rent, average education, job level, home value, and family income. Figures like these are frequently used as justification for urban renewal and slum clearance. However, a different image from that conveyed by census data is created by walking through the area. It is true that in El Barrio the housing is old and that excessive subdividing has created crowded conditions in some sectors. However, most buildings are structurally sound and well maintained, which makes the area a more desirable place to live than one would expect from its low rank in the city. The median income is $8,000 per year and an extremely low 2.3 percent of the population are unemployed. During the time of 1970 census was given, only 1.3 percent of the residents of El Barrio were on welfare.

These figures are considerably lower than the national average and present a striking contrast to welfare and unemployment rates for other ghetto populations.

The situation in the El Barrio community is one common to many inner city areas that have gone through rapid sociocultural change. In order to accommodate El Barrio's new Mexican population, city and community institutions have been under some pressure to change the nature of their services. This change is usually resisted, and even when changes do occur they are agoniz-

ing and frequently unsuccessful. The response of Mexican residents has been to avoid contact with these institutions and seek alternate resources among friends and relatives to meet their needs. The political and economic powerlessness of this group has allowed these institutions to continue to resist significant changes in policy and operation. This situation is particularly evident in the area of health services where unique health needs are not being met by standard American health institutions.

The conclusions of our own and other research efforts point to clear differences between Mexican-American and general American populations in disease rates, health attitudes and practices, disease configurations, and psychopathology. These differences can be summarized as follows:

1. Chicano death rates are higher than national averages as a consequence of the diseases of poverty including influenza, pneumonia, tuberculosis, neonatal death, and rheumatic fever. Alcoholism and drug addiction, traumatic injury, and infectious conditions exacerbated by malnutrition are also recognized as major problems.

2. Unlike Anglo-American medical beliefs, the traditional Mexican view of disease causation and symptomatology does not reflect a distinction between the mental and physical aspects of healthy functioning; this interrelationship between physical and mental factors is a key to the lack of understanding of Mexicans toward the separation of medical and psychiatric services.

3. Mexican-American beliefs about disease causation and symptomatology include a large number of illnesses that are unique to Mexican-Americans as a group. These disease configurations, or "folk illnesses," include **el ojo malo** [the evil eye], **empacho** [stomach upset], **bilis** [the product

of extreme anger], and **susto** [result of fright or shock].

4. The existence of **curanderas** or traditional Mexican medical practitioners is a very important health resource in the El Barrio community. From our research we can say with some confidence that more individuals utilize folk curers in El Barrio than use "standard" medical and psychiatric facilities. The **curanderas** provide low-cost care with an emphasis on a personalized and "sacred" approach. They use a large herbal inventory in addition to other techniques such as dietary restrictions, chiropractics, and religious-magical curing.

5. Mexican psychiatric patients show pathology and personality structures that are clearly different from those reported for other ethnic groups. For example, visions and voices are a widely acceptable part of normal functioning and maintenance of health in the Chicano population. A member of our research team has demonstrated the potential for misinterpreting visions and voices as psychopathology by non-Latin psychiatric staff[84].

6. Our research in the El Barrio community indicates considerable ethnic and intraethnic diversity in other health-related problems. In drug addiction, for example, we find that patterns of drug use and life situations among Chicano addicts are quite different from what has been described by addicts of other ethnic groups. We have also observed differences in drug use and life situation that distinguish first-generation Mexican addicts from those who have come to Chicago from Texas and those that were born in Chicago[85]. Other research currently being conducted in the El Barrio community on psychiatric difficulties, alcoholism, and old age is beginning to indicate a wide range of additional factors specific to this ethnic group.

The reluctance of Mexican-Americans to utilize medical and psychiatric facilities has been well documented. This underutilization involves a number of factors including: the existence of **curanderismo,** the tendency of American doctors to scoff at folk beliefs, as well as outright rudeness and racism on the part of health professionals. Madsen[86], Clark[87], and others, have made recommendations for professionals concerning possible changes in their behavior and procedures to bring them into closer fit with the standards of Chicano culture. These and other studies (e.g., [88]) document instances in which Chicano patients, or would-be patients, have been discouraged by what seems to them to be a rejecting or patronizing attitude on the part of medical personnel.

From the standpoint of first-generation immigrants, Chicago's El Barrio community may be among those most culturally and linguistically Mexican in the United States. A continuing flow of new migrants promises to maintain this strong Mexican orientation. As a consequence, we can expect that many of the health attitudes and practices described above will continue to be salient to this community, rather than diminish in importance as has been noted in other Mexican communities.

THE COMMUNITY MENTAL HEALTH PROGRAM

The Community Mental Health Program, through the El Barrio Mental Health Clinic, represented the first publicly funded service to direct its attention to an aspect of health in the Mexican community. However, given El Barrio's broad health needs and the lack of information about the functions of the mental health outpost, the Mexican-American people did not consider the program to be relevant to their health needs. In addition, the program's narrow definition of mental health care precluded a broad attack on the community's health problems.

In its first several months, the clinic found itself devoting most of its time to serving older middle-European patients who had long histories of mental illness and hospitalizations. Few Mexican residents sought help at the clinic and on the whole its existence and services were large ignored. The difficulties in establishing an effective mental health service in the El Barrio community were exacerbated by the fact that the clinic was staffed almost exclusively by non-Latins who lacked an understanding of Mexican culture and the necessary bilingual ability for communicating effectively with Spanish-speaking patients. In addition, the clinic was burdened by a series of bureaucratic and political contingencies in the medical center that made it difficult for the various components of the Community Mental Health Program to coordinate their services effectively.

Another problem faced by the program in this area was that of citizen participation. An important part of the community mental health movement was the involvement of citizens and consumers in the direction and formulation of program policies and objectives. Several attempts to form advisory boards for the clinic failed, and the idea of advisory boards was eventually abandoned. Thus, residents neither used the services nor were very interested in becoming involved in the clinic's operations. The clinic, and in turn the program, stood outside of the mainstream of community life and only tangentially related to the community's health needs.

There were great expectations on the part of the clinical staff that our Community Research Unit could quickly discover the key cultural and community factors that could solve their problems in communication, underutilization, and citizen participation. Because these problems were very real and immediate, the clinical staff felt that the anthropologists had to provide this information

almost immediately even though we had not been brought on the staff until one and one-half years after the clinic had been established. It soon became apparent to the clinicians that we had not entered the situation with a prepackaged set of principles that would immediately help them out of the difficulties they faced in the community. They also communicated to us that they were not willing to give us the time we needed to learn about the community and its people. Their reaction was to "write us off" as an important component of the program; as a result, there were the inevitable clinician-researcher tensions in our relationship to the rest of the program. After this initial interaction with the clinical staff it became clear to us that information about the community and its various cultural and ethnic groups was producing little interest and only minimal changes in the program. While in retrospect we can see that both sides failed to appreciate the point of view and professional concern of one another the effect at that time caused us to withdraw from intensive involvement in the clinical and policy aspect of the program. We turned to a search for new situations in which our research data could make useful contributions to positive social action. We found these action situations as a part of the process of community development that had already begun in El Barrio.

COMMUNITY DEVELOPMENT AND HEALTH SERVICES

The Rise of a Chicano Organization in El Barrio

In the initial stages of our involvement in El Barrio in 1969, one community organization dominated the scene. This organization, the Neighbors' Group, was established in 1954 and had its roots among the middle-Europeans in the community. The Neighbors' Group had developed a buying cooperative, housing and community-development committees, and a credit union. However, an increasing number of Mexican residents felt that the Neighbors' Group was not working effectively for Latin people though the group had made an effort to recruit Mexican members. The firing of several of the Mexican staff led to the development of a new group emphasizing its Mexican background and challenging the policies and programs of the older organization. The struggle between these organizations ended with the collapse of the Neighbors' Group, and the Chicano group began a broad-based attack on the problems of education, urban renewal, and a number of other major issues.

The strategy of our Community Research Unit in this period was to develop positive relationships with leaders and activists in all sectors of the community. Our newness to the situation allowed us to maintain these relationships without having to choose sides and to view community events as neutral observers. However, this "objectiveness" as well as the lack of identity in the area of the Community Mental Health Program served to keep us on the periphery of community life.

Organizing Residents' Groups

In June 1969 a significant breakthrough occurred in our relationships with community residents. The local settlement house developed a program in community organization in which clubs would be organized on scattered blocks throughout the neighborhood. Our decision to become intimately involved with this effort provided entree and rapport for our fieldworkers, and gave us the opportunity to demonstrate to an important sector of the population that the information we were able to collect and dissem-

inate could make a significant contribution to the goals of their programs.

Throughout the following year our tactics were to seek out opportunities in which our research personnel could be useful to community groups that were involved in a broad range of problems and issues. In this period, general ethnographic data, the results of survey operations on the blocks and the schools, and information collected through the program provided a body of material that proved to be useful to these groups. For example, information we collected through surveys of public and parochial school students and their parents proved useful to a community group working on education.

A Period of Organizational Decline

Toward the end of 1970, the Chicano group began to experience great difficulty in maintaining their objective of dealing with a wide variety of community problems. They had difficulty achieving success on individual issues and were overcommitted in a number of areas of community action. As a consequence, attendance at meetings began to decline and the organization lost a number of members. Over a period of several months, organizational activity declined significantly and, soon after, the organization existed in name only. Other community organizations with broad mandates also experienced difficulties in maintaining their efforts during this time. The decline in organizational activity was made more severe by the fact that block residents' organizations, supported by settlement house staff during the summer, never seemed to maintain themselves during the nonsummer months.

In this period of scattered and inconsistent effort in which community action groups shifted from one issue to another, all we could do was hope that we had appropriate bodies of data to address immediate concerns, or that we could construct a "rough and ready" operation for quick feedback. In this fluid situation prior to 1971, community action objectives were often unclear and efforts were transitory, and as a result we were frequently caught with insufficient information to contribute. Our strategy, up to this point was to:

1. Construct research operations in areas that we thought would have maximum benefit for community action research,
2. Seize opportunities in which community action groups, concerned residents, or agency staff could provide us with entree to data-gathering opportunities in the community,
3. Emphasize rapid feedback of research information to Chicano organizations and individuals in the El Barrio community,
4. Participate actively in community action organizations to the extent that such activity was approved by its members.

The Rise of Specialized Action Groups

In the beginning of 1971 a new climate began developing both in the wider society and in the local community. The apparent success of Cesar Chavez and the Farmworkers, the efforts of Tijerina in New Mexico and Corky Gonzales in Colorado, the increasing demand for recognition of Chicanos in the Midwest, and the developing sense of Chicano identity made people in the El Barrio community aware of the role they could play in a solution to community problems. At the same time a group of Chicanos who had gained organizing knowledge in past community efforts developed a number of voluntary groups, each of which focused on specific issues. These efforts centered around the creation of new youth facilities and greater availability of educational opportunities, and were linked to demands for significant bilingual and bicultural programs in area schools and equal opportunities in jobs.

A significant portion of these specialized activist groups began to direct their attention toward the health services situation in the community. For example, one group, with the help of some volunteer medical personnel, organized a free health clinic in a neighborhood settlement house to serve the El Barrio community. A group of Chicano ex-addicts developed a volunteer program in drug abuse, and a group of residents with experience and training in mental health sought ways to increase the number of Chicano mental health workers in the area. These efforts were organized for the most part by Chicano residents who had special experience, training, and talents for working on these health issues. Our applied research group was fortunate in this period to have established good working relationships with many of these groups. As they were getting organized we were asked to participate in a number of aspects of the groups' development and at the same time to contribute research results on some of these specific issues.

With the development in 1971 of this new stage in which community organizations established long-term and relatively concrete goals in specialized areas, we began to plan more long-term and specific operations that could have direct input into the efforts of these groups. The first goal of these specialized action groups was to make institutions both in the community and in the city more relevant to Chicano populations. For example, through the efforts of a group of community activists more bilingual and bicultural personnel were added to the staff of the El Barrio Mental Health Clinic.

However, attempts in other areas met with considerably less success. A group of Chicano ex-addicts sought help from the state drug program for the drug abuse situation in the El Barrio community but the state was unresponsive. A number of groups met the same reception from city, federal, and private agencies. As it became evident that existing agencies were not going to change to meet the needs of the El Barrio residents, these community action groups shifted from their initial institutional change objectives and began working toward the development of alternative community-run facilities. Several of these groups sought federal and private funding to establish such programs. The problem now became one of gaining funding and planning alternative programs. Community action strategy shifted to one of trying to take advantage of some of the money that was newly freed for "minority" community development. This search for outside funding was greatly enhanced by the effects of the Chicano movement in the wider society as well as in the community. The movement had put increasing pressure on federal agencies to provide funds to Chicano groups. Research in a number of fields had shown that there were special requirements in the areas of health, mental health, and social service programs for Chicano populations. The federal administration had begun to view Chicano groups as an increasingly potent force in determining the outcome of national and state elections. Therefore, there were political motives favoring the granting of funds to Chicano groups.

Community organizations and activists were well-prepared to take advantage of this opportunity. By 1972, they had gained experience in confronting institutions, had developed a good working knowledge in their special areas of concern, and had built effective organizations with considerable community support.

For the first time, we had a very specific context in which our research could be utilized: "grant proposals," i.e. documents for which we supplied extensive portrayals of community problems and their effect on residents as part of the background, rationale and justifications for requests for federal, state and private funds. The research data collected in our initial two years were reexamined in the light of the informational needs of each funding proposal. In addition, we were able to conduct interviews and examine archival data to help members of each group to express their ideas within the specific format of the proposals.

The procedure of submitting proposals involved a very close collaboration between members of our research team and community activists. The relationships established in the course of working on these proposals were to be extremely important for later and more extensive collaboration between our Community Research Unit and specific health programs.

Our data at this point were not yet extensive enough to provide community people involved in the situation with new information they did not already have. But our statistics and qualitative descriptions were good enough to give outside funding agencies a clear picture of the community and its needs. At the same time, as these new programs were being formulated, it was becoming clear to many of the community activists that effective information-gathering procedures needed to be built into the development of their programs. By the time the proposals were submitted, special research operations had been developed on a collaborative basis between researchers and activists in order to do a better job in getting accurate information for the planning of new programs in the future.

In the period between June 1971 and June 1972 proposals submitted by community groups in El Barrio for children's programs included a comprehensive youth program, a day-care center, an outdoor recreational facility and a vocational program directed toward dropouts in the community. Proposals in the area of education included a bilingual library, a bilingual reading program and a college recruitment program for Chicano youth. In the area of health, proposals were submitted for the support of a program to train Chicano residents in mental health, for the support of a free-health clinic, and for the development of a Chicano drug abuse program to serve El Barrio residents.

Of the ten major proposals that had been submitted, only one was denied funding. In the fiscal year 1971–1972 there was no community-controlled federal and private-agency money in the El Barrio area, whereas in the current fiscal year (1972–1973) there is over $600,000. Grants for succeeding years now total over two and a half million dollars.

Establishment of Community Health Programs

1. The health services picture for the El Barrio community showed a considerable change as a result of this funding. There is now a free health clinic supported by volunteer help and federal funds, which provides health services to over 300 patients per week. This clinic has been able to establish a working and referral relationship which eases the problems of access and entree of community residents into the County Hospital and other facilities in the nearby medical center complex of West Side Chicago.

2. The Chicano Drug Abuse Program was able to expand from a small volunteer staff providing methadone and minimal counseling of thirty Chicano clients, to a comprehensive drug program providing a wide range of services to one hundred

clients. This program, operating with a bilingual and bicultural staff, is attempting to develop treatment modalities attuned to the culture and experience of Chicano clients and has been able to draw in clients who have never received treatment.

3. The Chicano Training Program is training thirty Chicano residents to become mental health practitioners. It has established a mental health curriculum leading to a A.A. degree taught by Chicano faculty and accredited by a local university. In the process of finding jobs for trainees, this program has brought about significant changes in hiring policy among mental health institutions, has developed greater sensitivity to Chicano mental health needs on the part of mental health institutions, and is formulating a new model of Chicano mental health services that will be more effective and culturally relevant for El Barrio residents.

4. The El Barrio Mental Health Center has become, over the past two years, an increasingly valuable resource in the overall health service picture of the community. The change was due directly to the fact that a recent Chicano social work graduate with long experience in the community was hired as the director. He very quickly moved to recruit Spanish-speaking personnel, opened paraprofessional positions for residents knowledgeable about the community, and began to tie the clinic into other community-run service programs. The changes in policy are graphically reflected in the nature of the caseload at the clinic. During the period that these changes were occurring, the caseload shifted from approximately 80 percent middle-European and 20 percent Mexican in 1969, to 80 percent Mexican and 20 percent middle-European in 1972, which is reflective of the current population ratio[89].

There are close working relationships among the key staff people in each of these four programs. On many occasions, individual staff have been called upon to lend their expertise to mobilize community support, to use their contacts in both the local community and the wider society, and to provide other general support for each other's programs. The Chicano activist groups are now in constant communication with each other in developing new strategies, sharing information and consultation, and lobbying in the community's interest on a broad variety of issues. Because of this close collaboration, it became possible to coordinate services and referrals so that an effective health service system could be developed.

The building of this coordinated network has proceeded most effectively among those programs offering health and health related services. For example, in the area of services, the El Barrio Mental Health Clinic has detached a worker to the Free Health Clinic who can provide counseling to individuals whose difficulties have an emotional basis. The Drug Program utilizes both the Free Clinic and the Mental Health Clinic in dealing with the physical and emotional problems of drug addicts, and the Mental Health Clinic refers clients with drug problems to the Drug Program. The Chicano Training Program selected a number of its interns from the staffs of health and mental health agencies so that they could benefit from in-service training in a bicultural mental health program. In turn, Latin psychiatrists and psychiatric social workers on the staff of the Mental Health Clinic are part of the faculty of the Training Program, and have also provided specialized training to the staff of the Free Health Clinic and the Drug Program.

In addition to collaboration in the areas of education, training, and service there have been some meetings recently devoted to considering a coordinated research project that would satisfy much of the informational needs of all the social service programs.

While such an effort has yet to be fully established, our unit has made every attempt to stimulate its development so that the expertise to conduct needed research would exist in El Barrio.

Apart from defining common research interests, each program is now actively concerned with producing information that is relevant to the provision of services and justifies renewed funding. Such information includes survey data on the population served, evaluation of success criteria, and operations which examine new aspects of the problem in the general community. We are working closely with the personnel of these programs to help them provide this kind of information. In those programs where there are no research personnel, we have played the role of being the primary collectors of information. Thus, we now find ourselves in a much more difficult phase of research—that of attempting to furnish information of immediate practical utility to people involved in service programs.

In defining the specific research operations, we have made the most progress with those programs connected with the health service network described above. In response to the need for success criteria for clients in the Chicano Drug Program, we had developed a detailed interview schedule for gathering information on language use, origin, family and household structure, drug use, prison record, coping procedures, and income. A record-keeping procedure has been set up which will allow consistent data collection. We have interviewed staff members extensively concerning program goals, organization and attitudes. We are also conducting participant observation among addicts in the program and will soon be developing procedures to assess the nonclient addict in the community. Among other uses, this information was particularly useful in determining whether or not the

Drug Program was doing a good job in reaching all segments of the addict population in El Barrio. Finally, we want to compare the data collected on Chicano addicts with those of addicts from other ethnic groups to determine the unique features of Chicano addiction. These research and information collection procedures have been developed and are conducted in collaboration with staff of the Drug Program.

In the Chicano Training Program, our research team works closely with a part-time research assistant on the staff. Procedures have been developed for assessing the impact of the program on training skills, attitudes, behavior, and work performance, as well as in assessing the Training Program's impact in creating institutional changes. Within this program, work continues on an examination of **curanderismo** and its implications for the development of new Chicano mental health models.

In collaboration with the Free Health Clinic, a detailed survey instrument was developed to examine the health status, attitudes, and practices of residents in the El Barrio community. This instrument was seen as an important tool both in stating the case about health needs to outside groups and in providing information for the establishment of particular kinds of health programs at the free clinic. This survey will also be implemented by another free clinic in a Latin neighborhood to the west of El Barrio.

The developments over the past two years at the El Barrio Mental Health Clinic have resulted in closer cooperation between our unit and this component of the Community Mental Health Program. One result of this better relationship has been an increasing collaboration with psychiatric staff on research directed toward the role of social and cultural factors in mental and emotional difficulties in El Barrio. Currently, we are examining patient records and have drawn a

sample of clients from the clinic for intensive interviewing with a battery of interview schedules and psychiatric questionnaires.

This article has dealt with the development of health services and research in the Chicano community of El Barrio. In considering both service and research, we feel that the information we have presented here supports the view that change-oriented programs need to fit within the cultural and social life of the community and must be primarily determined and developed by community residents sensitive to indigenous behavior and concerns.

The experiences of our unit in working with the Mental Health Clinic particularly brings out this point. The clinic began its activities with models derived from Anglo-American psychiatric practices and was staffed by non-Latins who were unfamiliar with Chicano culture and community life. It demanded that the unit answer questions posed in terms of that alien model of psychiatric treatment. We felt, as anthropologists, that questions derived from community residents and activists took priority in terms of the information required to make mental health services more effective in the community. Thus, we turned away from our original task and directed our research to the informational needs of the developing indigenous human service system. As for the clinic itself, it was only when more bilingual and bicultural staff were hired that it sought to fit into the indigenous system, rather than insisting that it began to do an effective job in picking up a part of the responsibility for the establishment of a comprehensive health care delivery system in the community. Thus, a service and a research unit, both supported by state and federal funds, found themselves to be considerably more effective when they began to support and facilitate indigenous community organizations in achieving increased community development and in transforming their ideas into concrete community programs for human service. It is only now that we have reached this point—each being articulated to similar pressures and community events—that we have begun to realize a series of effective anthropological-psychiatric collaborations based on problems of health care delivery in the community context.

The establishment of this community health system, in our view, underscores two major generalizations. First, the development of adequate health services for "minority groups" and "economically disadvantaged" communities is a direct function of sociopolitical power. It is clear to us that had it not been for the Chicano movement and local successes in other types of programs, the El Barrio community would not have obtained the funds for the health programs that are currently operating. In turn, it is interesting to note that changes in health institutions and agencies serving El Barrio were impossible to effect until community groups were "legitimized" by receiving governmental and private funds. Now, at least in some sectors, these changes seem to come with relative ease and rapidity. We have learned, as a result, that information no matter how accurate, well-timed, and directed, always takes second place to political power in creating positive social change.

Secondly, the various health-related projects in El Barrio have established a coordinated effort in which organizations with different resources, expertise, and interests work together to create an optimal system from which to meet community needs. This coordination arises from close relationships among community activists and in turn their relationships with community residents. This collaborative arrangement contrasts strongly with the typical human services pattern in which a series of city, state, and federal agencies compete with each other and duplicate services.

It is important to note that the health programs in the El Barrio community still fall far short of meeting the total health needs of its residents. Specialized programs for alcoholics and senior citizens are still lacking and the Free Health Clinic, even at full capacity, serves only a small segment of the community population. While indigenous programs in El Barrio are working toward full coordination and cooperation, there are still ways in which this system could be improved.

However, the most overwhelming deficiencies in the health service system come as a result of the failure of federal, state, and local agencies to make funds available to the El Barrio community. While some improvement has been made in health care delivery, the medical establishment has a great many changes to make in staff and programs before they can claim that El Barrio residents have access to health care on a par with residents of other Chicago communities. Most community leaders and activists view the current situation as a beginning in the process of gaining their share of health resources for the community. They also clearly feel that unrelenting pressure must be applied if they are to make gains in the future.

The Planning Council and Neighborhood Health Care

By Robert R. Alford

Reprinted from *Health Care Politics* by Robert R. Alford by permission of The University of Chicago Press. Copyright 1975 by The University of Chicago Press.

The second illustration of political influences is an excerpt from Robert Alford's book Health Care Politics. *In summarizing his study of neighborhood health care centers, he describes how the impact of the diverse uncoordinated, and even contradictory, federal agencies and funding mechanisms further fragmented the network of health care agencies in the City of New York. Neither this nor the previous selection was conducted within the discipline of organizational research. Schensul and Bymel are anthropologists; Alford is a political sociologist. We are reminded of Aldrich's suggestion that conceptualizing organizations as boundary-maintaining systems can provide explanations of power and authority in terms applicable to all disciplines.*

Alford conducted a systematic study of the efforts to create neighborhood family care centers in New York City by means of a combination of federal, state, city and private funds. Interorganizational relationships had to be established at the local level; sponsorship had to be established at the higher planning level. His summary and conclusions illustrate some of the consequences of the complex problems for local health agencies when federal commitment to programs and funding vacillates. Battles over administrative jurisdictions compounded the problems.

Accusations of lack of cooperation, unwillingness to compromise, failure to communicate, and inability to define goals were frequent and led to attempts to create new channels of communication, new sets of rules, a redefining of goals and new officials to be responsible for coordination. In the absence of political mechanisms which could enforce translation of stated public policy into administrative implementation, no group or component of the array of health organizations provided by the city had the power or resources to impose its own definition of goals. Seventeen city-funded centers were planned in 1965; by 1971 ground had been broken for one. The "Summary and Conclusions" of Alford's research describe the lack of careful planning, failure to consider existing programs, failure to research needs, and the political necessity of developing highly visible programs. In the final analysis, Alford suggests that the fundamental features of the political economy of our health care system is responsible for the costly inefficiency:

My general theoretical perspective is that health care institutions, whether described as "fragmented" or as "pluralistic," must be understood in terms of a continuing struggle between major structural interests operating within the context of a market society—"professional monopolists" controlling the major health care

127

resources, "corporate rationalizers" challenging their power, and the community population seeking better health care via the actions of equal-health advocates.[90]

In general, then, we have seen that the clear and comprehensive guidelines of the Planning Council concerning coordination of the proposed new ambulatory care facilities with other health services already in the same geographic area were not followed, either in the proposals themselves or in the subsequent review process, and were not regarded as of sufficient importance to warrant requiring revision and resubmission of the proposals. In no case of the reading of a proposal and the subsequent correspondence among the Planning Council, the State Department of Health, and the applying agency did the mechanisms of coordination and integration become a major or even minor bone of contention.

There seem to be several clear reasons for this lack of concern with mechanisms of coordination. First, given the sudden availability of federal funds under various programs and their accompanying pressures to act quickly, and given the danger that other agencies, cities, or even states would consume those funds, both the applying agency and the reviewing bodies—whether the Planning Council or the state—were not disposed to develop the comprehensive data necessary to document the way in which the new center would be coordinated with existing health care facilities. The best that could be done was to develop an internal program, required staff and physical facilities, and tentative financing.

Second, given the pressing needs for more ambulatory care facilities in poverty areas of the city, the latent assumption was that almost *any* additional health care services would be worth providing. Information on how the new facilities would be coordinated

with old ones was not considered worth the time and resources to develop and integrate into the proposal. Almost all parties were willing to assume that mechanisms of coordination could be worked out ad hoc, or after the fact, once the funds had been obtained and the facilities were in operation.

The justification offered by the Planning Council in ultimately approving plans by the city to build "prototype 55" ambulatory care facilities without specifically tailoring each program to the needs of the community was that **any** additional health services in poverty areas were worthwhile. In the absence of any empirical specification of those needs and an evaluation of the total network of health services available to the community, it seemed utopian or excessively rigid to require the city to prepare a plan which presupposed that kind of research and policy planning. The council itself had not conducted any research which laid down criteria for community needs and had not systematically assessed the available health services in the city.

Given the complexity of the system—the diversity of origins of new services, funding sources, and controlling agencies—it would be extremely difficult to compile even a list of the services provided by health care facilities in an area, let alone assess whether or not those facilities were meeting a reasonable definition of needs of the community. Thus, requiring the city to prepare a program tailored to the community's needs would establish a standard which presupposed a level of planning capability that did not exist either in the council or anywhere else. This lack of planning capability is part of the phenomenon to be explained, not an explanation of why "coordination" did not take place or why "fragmentation" continued to exist. To assert that the fragmented system existed **because** of a lack of proper research, knowledge, or communication is merely to illustrate still another feature of a fragmented system, not to discover a cause which explains the fragmentation.

Third, in the case of the Health Services Administration, there were the political necessities of developing a new, highly visible program and of showing some results. The cost of possibly including some unnecessary facilities in one or more of the "prototypes" may have seemed, quite plausibly, justified to the city health planners, both in terms of getting something—anything—built before building costs escalated any further, and in terms of getting some additional visible services, in the form of a tangible structure labeled a Neighborhood Family Care Center, out into politically sensitive neighborhoods during the tenure of the Lindsay administration in New York City.

This last point is related to a more general one: namely, that all agencies, whether directly "political" or not, have to justify their existence by new and better programs, and the question of how these programs will be integrated with existing ones is quite low on the priority list. In fact, from the point of view of the organizational incentives involved, it may be that the less integrated and more autonomous, isolated, and dependent the new staff and agency are upon the administrative or political officials which sponsored them, the more likely the agency is to provide legitimacy for the sponsoring agency or officials and thus to become part of the resources of their political and organizational base.

Because of the absence of any coordinating mechanisms for consolidating decision making and planning, the officials of each organization went ahead on their own, trying to maintain their institutions and to solidify their position vis-a-vis relevant publics and other funding or regulating agencies. Each organization, in developing a project proposal, attempted to develop support and move ahead to the point where it would be difficult for other organizations to stop it. This may be one of the factors accounting for communication and coordination problems. If full information were available to other organizations before or at the early stages of planning, funding, designing, and so forth, they would be better able to block the project. Sometimes the interests of other organizations are well known—or at least it is known which organizations are most likely to either favor or oppose a project—so that the "friends" can be contacted and mobilized to support the project and the potential "enemies" can simply not be informed until, hopefully, it is too late. This factor seems most likely to operate where resources are either scarce or vacillating.

The history of the Neighborhood Family Care Centers illustrates the extreme dependence of local health programs upon the vicissitudes of federal legislation. The rise of a particular "hot" program such as the War on Poverty generates a flurry of activity—plans, proposals, meetings, new organizations—but this activity quickly dies down as another program appears which is advertised as solving the problems. However, the "crisis" continues, largely untouched, because no program which is politically feasible can also attack the causes of the problem and more than a few of the consequences.

The process of applications for funds for and approval of new Neighborhood Family Care Centers thus cannot be evaluated on the basis of "rational" criteria. Funding and programs, and hence priorities, constantly change, and planning organizations must take into account these persistent conditions of the organizational environment. In 1965, neighborhood health centers were funded; in 1971, family planning and abortion. The impetus behind ambulatory care has dwindled, and as of 1971 there was almost no action in that area. Only one NFCC was under construction, and the prospects for more were gloomy.

In conclusion, it can be said that despite the attention given during this period to the development of neighborhood health centers as a solution to a pressing problem, the program, "even at the height of its proliferation in New York City . . . never . . . even accounted for more than 5 percent of the institutional ambulatory care visits here. Today this percentage has probably shrunk. The centers are struggling to survive in the face of dwindling financing. Also, the New York City 'Prototype 55' program seems to be moribund."[91]

As was the case with other federal programs, funds were available for construction but not for operating expenses. Demonstration projects were possible in a variety of areas (such as methadone clinics), but no continuing commitment was likely, regardless of the success or failure of a project. Thus a flurry of activity occurred as separate programs rose and fell, together with their implementing organizations. This process exemplifies another aspect of a "fragmented" health system. While a proximate explanation for a particular case can be sought in failures of leadership, inadequate funding, errors of foresight or planning, mistaken decisions, or lack of public pressure for action, an ultimate explanation is probably to be found in fundamental features of the political economy of health care in the United States.

Alford believes that there are three different models or theories of the causes of existing arrangements in health care. The pluralistic or market perspective accepts the struggle of interest groups as inevitable and even necessary. The second bureaucratic or planning perspective holds that it is possible nationally to plan and coordinate the health care delivery system. The third institutional or class perspective holds that the defects in the performance of health institutions are duplicated in many other areas of American society and the roots lie deep in the structure of a class society. Policy research such as Alford's increases one's depth of understanding about the larger health delivery system within which one's own organization is just a small part.

REFERENCES AND NOTES

1. de Greene, K.B. Organizational Best Fit. In Burbank, E., and Negandhi, A. (Eds.) *Organization Design: Theoretical Perspectives and Empirical Findings*. Kent, Ohio: Kent State University Press, 1977.
2. Evans, W.A. *Organizational Theory: Structures, Systems, and Environments*. New York: John Wiley, 1976, p. 259.
3. Osborn, R., and Hunt, J. Environment and organizational effectiveness. *Admin. Sci. Quart.*, 19:231–45, June 1974.
4. Richman, B., and Farmer, R. *Management and Organizations*. New York: Random House, 1975.
5. Van de Ven, A., Emmett, D., and Koenig, R. Frameworks for Interorganizational Analysis. In Negandhi, A. (Ed.) *Interorganization Theory*. Kent, Ohio: Kent State University Press, 1975.
6. Kimberly, J.R. Environmental constrants and organizational structure: a comparative analysis of rehabilitation organizations. *Admin. Sci. Quart.*, 20:1-10, March 1975.
7. Dill, W. Environment as an influence on managerial autonomy. *Admin. Sci. Quart.*, 2:409-43, 1958.
8. Dill, W. 1958.
9. Osborn, R., and Hunt, J. 1974.
10. Kinzer, D.M. *Health Controls out of Control*. Chicago: Teach'em, 1977.
11. Kinzer, D.M. 1977, p. 139.
12. Richman, B., and Farmer, R. 1975, p. 80.
13. Lorsch, J.W. Environment, Organization, and the Individual. In Negandhi, A. (Ed.) *Modern Organizational Theory*. Kent, Ohio: Kent State University Press, 1973, p. 132.
14. Lorsch, J.W. 1973, p. 134.
15. Burns, T., and Stalker, G.M. *The Management of Innovation*. London: Tavistock, 1961.
16. Lawrence, P.R., and Lorsch, J.W. Organization and Environment. *Managing Differentiation and Integration*. Homewood, Ill.: Richard D. Irwin, 1967.
17. Child, J. Organization Design and Performance: Contingency Theory and Beyond. In Burack, E., and Negandhi, A. (Eds.) *Organizational Design: The Theoretical Perspectives and Empirical Findings*. Kent, Ohio: Kent State University Press, 1977.
18. Melcher, A., and Oberg, W. Comments on Lorsch's paper. In Negandhi, A. (Ed.) 1973, pp. 198-99. Child, J. 1977, p. 173. Evans, W.A. 1976, p. 262. Osborn, R., and Hunt, J. 1974, p. 233.
19. This research was supported in part by the Institute of Labor and Industrial Relations at the University of Illinois and in part by a grant from the Rehabilitation Services Administration, Department of Health, Education and Welfare. The author would like to acknowledge the support of the Center for Advanced Study at the University of Illinois in making writing time available through a Resident Fellowship. Helpful comments on an earlier draft of this paper were provided by William Button, William Form, James Garrett, James O'Connor and Jeff Pfeffer.
20. Bendix, R. *Work and Authority in Industry*. New York: John Wiley, 1956.
21. Abegglen, J.C. *The Japanese Factory*. Glencoe, Ill.: Free Press, 1958.
22. Crozier, M. *The Bureaucratic Phenomenon*. Chicago: University of Chicago Press, 1964.
23. Stinchcombe, A.L. Social Structure and Organization. In March, J.G. (Ed.) *Handbook of Organizations*. Chicago: Rand McNally, 1965, p. 153.
24. Hall, R.H. *Organizations: Structure and Process*. Englewood Cliffs, N.J.: Prentice-Hall, 1972, pp. 306–12, esp. p. 310.
25. Burns, T. The Comparative Study of Organization. In Vroom, V.H. (Ed.) *Methods of Organizational Research*. Pittsburgh, Pa.: University of Pittsburgh Press, 1967, p. 133.
26. National Association of Sheltered Workshops and Homebound Programs. *Sheltered Workshops—A Handbook*. Washington, D.C.: NASWHP, 1966.
27. Kimberly, J. The Financial Structure of Sheltered Workshops. Region II Rehabilitation Research Institute Report Series, No. 3, 1968.
28. Salmon, P.J. Historical Development of the Special Workshop: Its Present Employment Training and Placement Practices. In U.S. Department of Health, Education and Welfare. *Workshops for the Disabled*. Rehabilitation Services Series, N. 371. Washington, D.C.: DHEW, 1959, pp. 66–97. Olshansky, S. The transitional sheltered workshop: a survey. *J. Social Issues,* 16:203-13, 1960.
29. Wessen, A.E. The Apparatus of Rehabilitation: An Organizational Analysis. In Sussman, M.B. (Ed.) *Sociology and Rehabilitation*. Washington, D.C.: American Sociological Association, 1965, pp. 141-63.
30. Wessen, A.E. 1965, p. 149.
31. Wessen, A.E. 1965, p. 149.
32. Olshansky, S. 1960, p. 33.
33. Olshansky, S. 1960, p. 74.
34. Wessen, A.E. 1965, pp. 145-50.
35. Wessen, A.E. 1965, p. 1.
36. Wessen, A.E. 1965, p. 151.
37. Thompson, N.Z. *The Role of the Workshop in Rehabilitation*. Bedford, Pa.: National Institute on the Role of the Workshop in Rehabilitation.
38. $n = 123, df = 1$
$\chi^2 = 24.09 p < .001$
$r_o = .42 p < .01$
zero-order $r = .44 p < .01$
point biserial $r = .37 p < .01.$

The zero-order correlation was computed between

the number of years each organization had been in operation and the percentage of total income it derived from production activities. The point biserial correlation was computed between the number of years each organization had been in operation and organization orientation, a dichotomous variation.

39. Stinchcombe, A.L. 1965, pp. 153–64.
40. Stinchcombe, A.L. 1965, p. 161.
41. Thompson, J.D. *Organizations in Action.* New York: McGraw-Hill, 1967.
42.
$$n = 123, = 1$$
$$\chi^2 = 13.91 p < .001$$
$$r_\phi = .40 p < .01$$
$$\text{zero-order } r = .44 p < .01$$
$$\text{point biserial } r = .36 p < .01.$$
43. Blau, P.M. and Schoenherr, R. *The Structure of Organizations.* New York, Basic Books, 1971.
44. Aldrich, H. An Organization-Environment Perspective on Cooperation and Conflict between Organizations in the Manpower Training System. In Negandhi, A. (Ed.) *Interorganization Theory.* Kent, Ohio: Kent State University Press, 1975.
45. Richman, B., and Farmer, R., 1975, p. 78.
46. Evans, W.A. 1976.
47. Aday, L., and Eichhorn, R.L. The Utilization of Health Services: Indices and Correlates. A Research Bibliography. Publ. No. (HSM) 73-3003. National Center for Health Services Research and Development. Washington, D.C.: U.S. Government Printing Office, 1972.
Anderson, J.G. Demographic factors affecting health services utilization: a causal model. *Med. Care,* 11:104, March-April 1973.
Anderson, O.W., and Anderson, R.M. Patterns of Use of Health Services. In Freeman, H.E., Levine, S., and Roeder, L.G. (Eds.) *Handbook of Medical Sociology.* Englewood Cliffs, N.J.: Prentice-Hall, 1972, pp. 386–406.
Anderson, R.M., and Newman, J.F. Societal and individual determinants of medical care utilization in the United States. *Milbank Mem. Fund Q/Health Soc.,* 51:95, Winter 1973.
48. Wirick, G.C., Jr., Morgan, J.N., and Barlow, R. Population Survey: Health Care and Its Financing. In McNerney, W.J. et al. (Eds.) *Hospital and Medical Economics,* vol. I. Chicago: Hospital Research and Educational Trust, 1962.
49. Anderson, J.G. 1973.
50. Anderson, R.M., and Anderson, O.W. *A Decade of Health Services.* Chicago: University of Chicago Press, 1967.
Anderson, O.W., and Sheatsley, P.B. *Hospital Use: A Survey of Patient and Physician Decisions.* Research Series, No. 24. Center for Health Administration Studies. Chicago: University of Chicago Press, 1967.
Bashshur, R.L., Shannon, G., and Metzner, C. Some ecological differentials in the use of medical services. *Health Serv. Res.,* 6:61, Spring 1971.

Durbin, R.L., and Antelman, G. A study of the effects of selected variables on hospital utilization. *Hosp. Manage.,* 98:57, Aug. 1964.
Fitzpatrick, R.B., Riedel, D.C., and Payne, B.C. Character and Effectiveness of Hospital Use. In McNerney, W.J. et al. (Eds.) 1962, pp. 361–591.
Greenlick, M.R. et al. Determinants of medical care utilization. *Health Serv. Res.,* 3:296, Winter 1968.
Wilder, C.S. Hospital Discharges and Length of Stay: Short-Stay Hospitals: United States—July 1963–June 1964. PHS Pub. No. 1000. Vital and Health Statistics, Series 10, No. 30. Washington, D.C.: U.S. Government Printing Office, 1966.
Witkin, M.J. Utilization of Short-Stay Hospitals by Characteristics of Discharged Patients: United States—1965. PHS Pub. No. 1000. Vital and Health Statistics, Series 13, No. 3. Washington, D.C.: U.S. Government Printing Office, 1967.
Ro, K., Patient characteristics, hospital characteristics, and hospital use. *Med. Care,* 7:295, July-Aug. 1969.
Rosenthal, G.D. *The Demand for General Hospital Facilities.* Hospital Monograph Series, No. 14. Chicago: American Hospital Association, 1964.
Wirick, G.C., Jr. A multiple equation model of demand for health care. *Health Serv. Res.,* 1:301, Winter 1966.
51. New York State Office of Statistical Coordination. *New York State Statistical Yearbook—1971.* Albany, N.Y.: New York State Division of the Budget, 1971.
52. Fitzpatrick, R.B., Riedel, D.C., and Payne, B.C. 1962. Ro, K. 1969. Skinner, C.G. *Hospital and Medical Economics,* vol. 2. Chicago: Hospital Research and Educational Trust, 1962, pp. 717–868.
53. Fitzpatrick, R.B., Riedel, D.C., and Payne, B.C. 1962.
54. American Hospital Association. Listing of hospitals. *Hospitals,* 45, Aug. 1, 1971, Part 2, Guide Issue.
55. Harris, D.M. Some Determinants of the Diversity and Local Concentration of Hospital Environment Interorganization Relations. Ph.D. thesis, State University of New York at Stony Brook, 1973.
56. Fitzpatrick, R.B., Riedel, D.C., and Payne, B.C. 1962.
Richardson, A.M. et al. Use of medical resources by SPANCOS II: Social factors and medical care experience. *Milbank Mem. Fund Quart.,* 45:61, Jan. 1967.
Roth, J.A. The necessity and control of hospitalization. *Soc. Sci. Med.,* 6:425, Aug. 1972.
Shannon, G.W., Bashshur, R.L., and Metzner, C.A. The concept of distance as a factor in accessibility and utilization of health care. *Med. Care Rev.,* 26:143, Feb. 1969.
Weiss, J.E., and Greenlick, M.R. Determinants

of medical care utilization: the effect of social class and distance on contacts with the medical care system. *Med. Care,* 8:456, Nov.-Dec. 1970.

57. Anderson, J.G. 1973. Durbin, R.L., and Antelman, G. 1964. Skinner, C.G. 1962. Roemer, M.I., and Shain, M. *Hospitalization Utilization under Insurance.* Hospital Monograph No. 6, Chicago: American Hospital Association, 1959.

58. Rosenthal, G.D. 1964.

59. Pelz, D.C., and Andrews, F.M. Detecting causal priorities in panel study data. *Am. Sociological Rev.,* 29:836, Dec. 1964.

60. Harris, D.M. An elaboration of the relationship between hospital bed supply and general hospital utilization. *J. Health Soc. Behav.,* 16:163, June 1975.

61. Ro, K. 1969.

62. American Hospital Association, 1971.

63. Aday, L., and Eichhorn, R.L. 1972.

64. Harris, D.M. 1973.

65. Duncan, O.D. Path Analysis: Sociological Examples. In Blalock, H.M., Jr. (Ed.) *Causal Models in the Social Sciences.* Chicago: Aldine-Atherton, 1971.
    Alwin, D.F., and Hauser, R.M. The decomposition of effects in path analysis. *Am. Sociological Rev.,* 40:37, Feb. 1975.
    Blalock, H.M., Jr. Evaluating the relative importance of variables. *Am. Sociological Rev.,* 26:866, Dec. 1961.

66. Duncan, O.D. 1971.

67. Ro, K. 1969.

68. Rosenthal, G.D. 1964.

69. Shain, M., and Roemer, M.I. Hospital costs relate to supply of beds. *Mod. Hosp.,* 92:71, April 1959.

70. Roemer, M.I. Bed supply and hospital utilization. *Hospitals,* 35:36, Nov. 1, 1961.

71. Klarman, H.E. Approaches to moderating the increases in medical care costs. *Med. Care,* 7:175, May-June, 1969.

72. Anderson, R.M., and Newman, J.R. 1973.

73. Harris, D.M. 1975.

74. Rogatz, P. Excessive hospitalization can be cut back. *Hospitals,* 48:51, Aug. 1, 1974.

75. Richman, B., and Farmer, R. 1975, p. 83.

76. Benson, J.K. The interorganizational network as a political economy. *Admin. Sci. Quart.,* 20:229–49, June 1975.

77. Aldrich, H. 1975, p. 51.

78. Richman, B., and Farmer, R. 1975, p. 83.

79. Schensul, S., and Bymel, M. The Role of Applied Research in the Development of Health Services in a Chicano Community in Chicago. In Ingman, S. and Thomas, A. (Eds.) *Topias and Utopias in Health.* Chicago: Aldine, 1975.

80. Benson, J.K. 1975.

81. While the major responsibility for writing this paper was ours, the development of the ideas was a collaborative process of community activists and members of our research staff. These included: Philip Ayala, Albert Vazques, Juan Velazquez, Humberto Martinez, Emile Schepers, Pertti J. Pelto, Elias Sevilla-Casas, Santiago Boiton, Susan Stechnij, and Kay Guzder.

82. Schensul, S.L. Preliminary Results from a Pilot Research Project in an El Barrio Elementary School. Unpublished, 1970.

83. Chicago Sun-Times, October 22, 1972, section la, p. 3

84. Schepers, E. Voices and Visions in Chicano Culture: Some Implications for Psychiatry and Anthropology. Paper presented at the American Anthropological Association meeting, Toronto, Canada.

85. Schensul, S.L. A New Model for the Understanding of Drug Addiction in a Chicano Community. Paper presented at the American Anthropological Association meeting, Toronto, Canada.

86. Madsen, W. Society and Health in the Lower Rio Grande Valley. In Burma, J. (Ed.) *Mexican-Americans in the United States.* Cambridge, Mass.: Schenkman.

87. Clark, M. *Health in the Mexican-American Culture: A Community Study.* Berkeley: University of California Press.

88. Rubel, A. *Across the Tracks: Mexican-Americans in a Texas City.* Austin: University of Texas Press, 1966.

89. Vasquez, A. The Effects of a Change in the Cultural Orientation of a Community Mental Health Clinic. Paper presented at the American Anthropological Association meeting, Toronto, Canada.

90. Alford, R.R. *Health Care Politics.* Chicago: The University of Chicago Press, 1975, p. xiv.

91. Personal communication from Herbert Williams, Senior Planning Associate, Health and Hospital Planning Council, 18 September 1973.

UNIT III
ORGANIZATIONS' ACCOMMODATIONS TO ENVIRONMENTAL PROBLEMS

- **ACCOMMODATIONS AT THE BOUNDARIES OF ORGANIZATIONS**
- **COMPETITION, COOPERATION, COOPTATION, NEGOTIATION**
- **PLANNING ORGANIZATIONAL RESPONSE TO ENVIRONMENTAL PROBLEMS**

Units I and II have used models, conceptual schemes, and empirical studies to describe developments in organizational theory based on open-systems concepts. The emerging theory cuts across sociological, political, and management sciences as theorists take into account the cultural, social, economic, and political environmental factors which influence organizational behavior. The impact of these factors on the internal structure of the organization has been examined. The implications of organizational interactions have, to some extent, been identified. Yet the theory does little beyond recognizing patterns of relationships, and managerial strategies cannot be prescribed with any degree of confidence. Nonetheless, the research literature addresses the critical nature of the organization's relationship to its external environment and the need of organizations to accommodate to environmental contingencies.

This unit describes strategies which organizations use to reduce uncertainty, maintain viability, and improve their ability to acquire scarce and valued resources. Attention is turned to boundary roles and how these are used for internal and external control. To tie the emerging theory to the practicalities of nursing administration, two articles from nursing literature are offered to describe how nurse managers have produced positive institutional response to community demands and mutual interagency problems. The nurse managers assessed the community situation, recognized barriers to organizational effectiveness, and capitalized on the opportunities afforded by the social and political environment.

The final selection, a paper by Aaron Wildavsky, is used to describe factors in the macroenvironment of health care institutions which shape how those institutions function. The selections that precede Wildavsky's should enable nurse managers to relate these concepts of regionalization to their new understanding of the dynamics of organization-environment relationships.

10. Accommodations At The Boundaries Of Organizations

A concept integral to an understanding of organization-environment relationships is that of the boundaries of organizations separating organization from environment. The nature of these boundaries and the activities they encompass are discussed in the literature from many points of view. Excerpts from articles by several researchers are presented here to give the reader an understanding of how organizational boundary components affect the capacity of the organization to adapt to environmental pressures and constraints.

The open-systems paradigm as applied in modern organization theory alerts us to the fact that the source of many organizational problems lies outside the organization's boundaries. This implies that administrators cannot accurately assess a problem unless they give attention to variations both inside the organization boundaries and outside the organization boundaries. Furthermore, the organization-environment perspective requires that careful consideration be given to both sides of transaction relationship which spans this boundary.

The concept of the boundary of an open system is difficult to establish empirically, especially if you are discussing an organization that has satellites, or one which has diffuse membership like the American Nurses' Association. Howard Aldrich suggests a definitive method for identifying where an organization stops and its environment begins.

The crucial element that defines an organization is the fact that a distinction is made between members and nonmembers, with an organization existing to the extent that entry into and exit out of the organization are limited. Some persons are admitted, while others are excluded.[1]

Various members of an organization conduct specialized activities which are directly associated with the transactions which

139

cross organizational boundaries. Buckley speaks of these mediating activities as follows:

The various features of complex adaptive systems sketched so far—openness, information-linkage of the parts and the environment, feedback loops, goal-direction, and so forth—provide the basic conceptual elements that underlie the general features characteristic of systems referred to as "self directing" and "self organizing." These concepts all point to the fact that the behavior of complex, open systems is not a simple and direct function of impinging external forces, as is the case with colliding billiard balls of gravitational systems. Rather, as open systems become more complex, there develop within them more and more complex mediating processes that intervene between external forces and behavior. At higher levels these mediating processes become more and more independent or autonomous, and more determinant of behavior. They come to perform the operations of: (1) temporarily adjusting the systems to external contingencies; (2) directing the system toward more congenial environments; and (3) permanently reorganizing aspects of the system itself to deal perhaps more effectively with the environment.[2]

*Miller and Rice use the term **boundary control function** to describe the specialized regulation activities which occur at the boundaries of organizations:*

What distinguishes a system from an aggregate of activities and preserves its boundaries is the existence of regulation. Regulation relates activities to throughput, ordering them in such a way as to ensure that the process is accomplished and that the different import-conversion-export processes of the system as a whole are related to the environment.

In the analysis of systems of activities, two types of regulatory activity can be identified: monitoring and boundary control. Monitoring refers to intra-system regulatory activities which are different in kind from, and not directly related to, the controls activated at the boundaries of the system.

Regulatory activities that relate a system to activities in its environment occur at the boundary of the system and the environment, and control the import and export transactions across it. Boundary regulation is therefore external to the operating activities of the system. The important implication is that the boundary around a system of activities is not simply a line, but a region with two boundaries, one between the internal activities of the system and the region of regulation, and a second between the region of regulation and the environment. For this form of regulation, we use the term boundary control function.[3]

Miller and Rice go on to discuss boundaries that are permeable, using hospitals as an illustration:

Strict boundary controls are especially difficult to maintain in those systems that by their very nature have to be more open. Hospitals, for example, find it difficult to control entry into their accident wards or into other emergency services. In general, those institutions and professions that offer help of any kind, physical or spiritual, frequently find that either their intake or their output is intractable to control. Those who come for help tend to be accepted—however hopeless their case—and once admitted are frequently difficult to export.[4]

Evans[5] states that top executives and some staff specialists, such as sales, public relations, and house counsel, are more likely to be engaged in boundary-maintenance functions than are junior and middle executives. The business lunch is a boundary-maintenance function of enormous proportions in top management circles of our society.

James Thompson, known as one of the foremost sociological scholars of organization, discusses organization boundaries in terms of "boundary spanning units,"[6] which must be interdependent with organizations in the environment, not subordinate to them. Thompson's concepts are similar to those of the contingency theorists we considered in Unit II. Rushing and Zald[7] credit Thompson as being the first systematic contingency theorist, even though Thompson never used the term **contingency theory** *in his writing.*

Two excerpts of Thompson's writing are included in this volume because his propositions offer challenging conceptualizations of how boundary structures are used to reduce organizational uncertainty.

Organizational Rationality and Structure (an excerpt)

By James D. Thompson

Reprinted with permission from *Organizations in Action* by James D. Thompson, pp. 66–73. Copyright 1967, McGraw-Hill Book Company.

Structural divisions are established to delimit coordinative complexity. Positions and groups of positions cluster in a manner calculated to handle the most critical aspects of their interdependence. The technology sets the constraints around which the organization manipulates its variables, and technical rationality is maximized when the variables are under complete control of the organization.

Because organizations are always embedded in larger systems of action, some parts of the organization must be interdependent with organizations not subordinated to the organization, hence not subject to authoritative specification of permissible action. The crucial problem for boundary-spanning units of an organization, therefore, is not coordination (of variables under control) but **adjustment** to constraints and contingencies not controlled by the organization—to what the economist calls exogenous variables.

Organizations subject to rationality norms seek to isolate their technical cores from environmental influences by establishing boundary-spanning units to buffer or level environmental fluctuations. These responsibilities help determine the structure of input and output units.

VARIETIES OF ENVIRONMENTAL CONSTRAINTS

Those elements of the task environment to which the organization must adapt vary from organization to organization, and do not fit neatly with any of the typical distinctions among organizations.

Some governmental units, for example, must live only on the financial inputs provided by legislative units such as Congress or a city council, and have no option even to which nation, state, or city they will attach themselves. But other governmental units do not find themselves so constrained regarding financial inputs. Many are empowered to raise capital funds through bond issues. Some may be self-supporting by charging fees for services; these may find, on the other hand, that their imperative is to deal with a specified clientele. This latter imperative may be equally true of the "private" public utility.

Prisons typically have no option regarding their clientele, and in this respect they may be compared with governmentally financed mental hospitals. Private mental hospitals, on the other hand, may be quite selective in their admissions policy[8].

For a given organization, the nature of en-

143

vironmental constraints may change over time. Thus at an earlier period, the voluntary hospital was a charity institution for the dying; its location and often its clientele was specified by wealthy donors[9]. As the domain of the hospital was redefined by the development of medical technology and the rise of third-party payment arrangements such as Blue Cross, the voluntary hospital was no longer confined to a geographic site or clientele specified by a single wealthy donor. The textile mill at one time was obliged to locate at the souce of water power, but the development of transmission lines and new sources of power freed the textile organization from this constraint.

Some organizations face constraints on the sources of raw material inputs; the extractive industries such as iron, petroleum, or lumber are examples. For others, raw material inputs may be easily available in any location or may be feasibly shipped and stored, but manpower inputs may constitute imperatives; various kinds of research organizations, for example, find they must locate where the relevant skills have been assembled.

It should be emphasized that the nature of an environmental constraint is not perfectly correlated with ownership of the organization, although Elling and Halebsky's[10] study of community hospitals indicates that the source of sponsorship has significant impact on other aspects of their actions. We cannot say that because an organization is public, or government-sponsored, it has a particular kind of environmental situation; nor that the private organization such as the business firm has another. Nor is there a one-to-one correlation between varieties of organizational constraints and the role in society of the organization. The impact of task environment is more subtle.

If these traditional modes of classifying organizations are inadequate for the pur-

pose, on what basis may organizations and their environmental constraints be compared? Generally, we may say that organizations find their environmental constraints located in **geographic space** or in the **social composition** of their task environments. And if this is the case, we need ways of characterizing both dimensions.

Typically, geographic space is described in terms of **distance** between points within it, but organizations usually measure this distance in terms of **costs of transportation** or **costs of communication.** The iron industry, for example, must bring together ore and coal, which are found at different points in geographic space. When the transporting of one or the other becomes easier or less costly, the firm's freedom to select from among various geographic locations becomes greater. Similarly, the development of electronic data-processing networks has the practical effect of reducing distance between points of the regionally scattered organization and thus permits centralization of decision making where heretofore localized units exercised local discretion in the interest of speedy response to local fluctuations.

There are fewer conventions with regard to the description and measurement of the social composition of environments. What we need are ways of differentiating among the kinds of social environments faced by complex organizations—the individual members, aggregates of individuals, and organizations which constitute task environments. A variety has been suggested recently. Task environments have been characterized by March and Simon[11] as hostile or benign. Dill[12] distinguished task environments as **homogeneous** or **heterogeneous, stable** or **rapidly shifting,** and **unified** or (by implication) **segmented.** Coleman[13] in describing communities, considered organizational density.

For present purposes, dimensions dealing

with degree of homogeneity and degree of stability seem most useful. In constrasting two firms, Dill[14] found it helpful to characterize the task environment as (relatively) homogeneous or heterogeneous, indicating whether the social entitites in it were, for organizationally relevant matters, similar to one another. In comparison with one serving a small city, the public school in an upper-middle-class suburb, for example, might enjoy a relatively homogeneous task environment with respect to the school expectations held by taxpayers and parents.

In Dill's study, the Alpha firm faced a relatively homogeneous environment, Beta, the heterogeneous one. Most of Alpha's customers ordered all four product lines, twice a year, and at the same time. The firm dealt with a single union, and virtually all of the external groups it dealt with were Norwegian. In contrast, Beta was active in a variety of quite distinct markets, dealt with three unions, and with suppliers and other groups in various parts of Europe. The task environment for both was pluralistic, but Dill makes the important point that not only did the firms differ in terms of the number of groups impinging on them, but that within each category (competitors, suppliers, customers, and regulatory agencies) the task environments were relatively homogeneous in one case and relatively heterogeneous in the other.

The social composition of task environments may also be characterized on a **stable-shifting** dimension. Dill noted, for example, that Alpha's market had changed relatively little in fifty years; although population had grown, the Norwegian family remained the only important customer. Products and means of distribution were basically the same. For Beta, however, markets that had existed when the firm was founded had grown in some cases and disappeared in others; only two of the major work

activities were continuations of pre-World War I, the other eleven having resulted from the growth and diversification of the industries which used Beta's services, from rapid technological development of the equipment and processes in which Beta specialized, and from increases in the real prosperity of the Norwegian consumer. Although the shifts in Beta's case are described here in terms of what we have called domain, they undoubtedly reflect more day-to-day fluctuation for Beta than for Alpha. Chandler[15] has described the significance of day-to-day fluctuations in demand for the producers of perishable products like meat or fruit.

There are undoubtedly other dimensions of task environment which have a bearing on organization structure, but these two appear at this point to be quite crucial, and they have the important advantage that they can be applied to all types of organizations and task environments. We believe they will help us understand differences, at least gross distinctions, in organizational structures. Although both dimensions affect organizations simultaneously, we will first consider their consequences independently.

BOUNDARY-SPANNING STRUCTURES

If adjustment or adaptability is the hallmark of boundary-spanning components of organizations, we would expect that fact to be reflected in the number and nature of the units established to handle boundary-spanning matters. Generally, we would expect the complexity of the structure, the number and variety of units, to reflect the complexity of the environment. If organization structure is an important means of achieving bounded rationality, then the more difficult the environment, the more important it is to assign a small portion of it to one unit.

Proposition 1: Under norms of rationality, organizations facing heterogeneous task environments seek to identify homogeneous segments and establish structural units to deal with each.

This proposition is perhaps most dramatically illustrated by the organization which crosses national boundaries where environmental variations may be stark. Under these conditions, organizations tend to establish semiautonomous divisions based on region. Those organizations which dabble in foreign operations may simply have a foreign or nondomestic division; but when they become more deeply involved, they usually establish national or at least bloc entities[16].

We need not cross cultural boundaries to see this proposition in action. Public school systems divide themselves into elementary and secondary schools, and only in rare cases are these ungraded internally. General hospitals establish separate units for obstetrics, contagious diseases, and surgical and outpatient services; mental hospitals often establish separate units for various types of disorders or severity of problems. Universities create undergraduate and graduate divisions. Public-assistance agencies may be divided into different units for dispensing unemployment compensation benefits and aid to dependent children or the blind. Air conditioning manufacturers may create separate divisions to make and dispose of residential and commercial units. Transport firms create separate divisions for passenger and cargo traffic. Large multiproduct retail establishments, facing heterogeneity on the input side, create specialized buying units.

Proposition 2: Under norms of rationality, boundary-spanning components facing homogeneous segments of the task environment are further subdivided to match surveillance capacity with environmental action.

If the sheer volume of interaction between a boundary-spanning component and its task environment is great, even though homogeneous, we would expect the organization to find a means of subdividing that component. Often this may be done by area or region in geographic space, but it may be done by otherwise insignificant differences in social composition of the task environment, such as alphabet.

Community mental health centers serve geographical catchment areas determined by numbers of people served.

Surveillance capacity undoubtedly varies as a result of differences in data-collecting, transmitting, and processing devices; but at any specific time the available devices set conditions for the organization. The degree of stability of the task-environment segment is a further constraint.

For instance, hospitals have set procedures for scheduling elective surgery to prevent too high a workload.

Proposition 3a: The organization component facing a stable task environment will rely on rules to achieve its adaptation to that environment.

For instance, a hospital in a summer resort area will not allow employee vacations during the summer months.

As in the case of coordination, we believe that adaptation by rule is the least costly form and therefore preferred by organizations which are under pressure to be efficient.

Proposition 3b: When the range of variation presented by the task-environment segment is known, the organization component will treat this as a constraint and adapt by standardizing sets of rules.

Many hospital emergency units have standing physician orders for standard treatment routines.

The organization may say that when the task environment behaves in manner A, respond according to rule 1; and when the task environment behaves in manner B, choose rule 2[17]. It is under these conditions that the bureaucratic procedures[18] of categorizing events and selecting appropriate response rules become very important. But this manner of proceeding becomes unwieldly if the range of possible variations in task-environment behavior is great, for it entails a proliferation of sets of rules, and places heavy burdens on the organization's capacity to categorize—to judge which set of environmental constraints it faces at a particular point in time and space.

Proposition 3c: When the range of task-environment variations is large or unpredictable, the responsible organization component must achieve the necessary adaptation by monitoring that environment and planning responses, and this calls for localized units.

Triage is one method health institutions use to sort out a wide variation of walk-in patients. One person assesses which patients receive priority, and the patients are sent to the clinic which is specialized to treat the presenting symptom.

Now, the above propositions treated each dimension independently, and indeed we believe they are independent in their actions. Nevertheless, all organizations face task environments which are located simultaneously somewhere on the homogeneous–heterogeneous continuum and the stable–shifting continuum. We can consider the interaction of the two as in Exhibit 3-1.

Exhibit 3-1. Interaction of Two Continuums for Task Environments

	Stable	Shifting
Homo		
Hetero		

From our above propositions, we would expect the organization whose task environment is relatively homogeneous and relatively stable to be relatively simple in the structure of its boundary-spanning components. This organization would have few **functional divisions,** and if these were subdivided it would be into several similar departments or sections. This organization would rely primarily on standardized responses or rules for adaptation, the departments or sections would be rule-applying agencies, and administration would consist of rule enforcement.

For the organization facing a heterogeneous but stable task environment, we would expect a variety of functional divisions, each corresponding to a relatively homogeneous segment of the task environment, and each relying primarily on rules to achieve adaptation. These functional divisions might be further subdivided, on geographic or similar bases, into similar or uniform departments or sections and would be rule-applying agencies.

When the task environment becomes dynamic rather than stable, new complications arise for the organization. Standardized response rules are inadequate, for the organization faces contingencies as well as constraints. It must determine when and how to act, and its cues must be taken from the task environment.

If the task environment is dynamic but relatively homogeneous, the boundary-spanning component need be differentiated or subdivided only to the extent that its capacity to monitor the environment would be overextended. Since, by definition, the environment is otherwise homogeneous, we would expect sections or departments of this organization to be established by area in geographic space. Unlike the regional divisions established for the stable-environment case, however, the regional divisions for the dynamic environment will be less concerned with the application of rules than with the planning of responses to environmental changes. When the task environment is dynamic, the regional divisions will be **decentralized.**

Finally, we have the situation in which the task environment is both heterogeneous and dynamic. Here we would expect boundary-spanning units to be differentiated functionally to correspond to segments of the task environment, and each to operate on a decentralized basis to monitor and plan responses to fluctuations in its sector of the task environment.

The sum of our argument regarding the impact of task environment on the structure of boundary-spanning units may now be expressed. The more heterogeneous the task environment, the greater the constraints presented to the organization. The more dynamic the task environment, the greater the contingencies presented to the organization. Under either condition, the organization seeking to be rational must put boundaries around the amount and scope of adaptation necessary, and it does this by establishing structural units specialized to face a limited range of contingencies within a limited set of constraints. The more constraints and contingencies the organization faces, the more its boundary-spanning component will be segmented. This argument seems consistent with that of March and Simon, who "predict that process specialization will be carried furthest in stable environments, and that under rapidly changing circumstances specialization will be sacrificed to secure greater self-containment of separate programs[19].

James Thompson speaks to the changing nature of administration as our social organizations become more complex:

Organizational survival rests on the co-alignment of technology and task environment with a viable domain, and of organization design and structure appropriate to that domain. This is the basic administrative function. The timing of both adaptive and directive action is a crucial administrative matter. Administration must keep the organization at the nexus of several necessary streams of action and find the strategic variables that can be manipulated, so as to result in a viable coalignment[20].

Thompson insightfully adds that attaining a viable domain is a political problem. In recognition of the increased interdependence of our social organizations, he writes:

But in modern societies, it appears, we have passed the era in which control and coordination of technological activities were the central administrative challenge, into an era in which organizational rationality is the core of

administration, and the administration of multi-organization projects and activities is the central challenge. Whether we have or will gain the knowledge about the organizations that it takes to use and control them under conditions of extreme interdependence remains to be seen.[21]

Negandhi[22] remarks that one of the main assumptions underlying the study of organization-environment interactions is that the boundary between the organization and its environment is clearly delineated and thus easily identified, yet our knowledge about the concept of boundaries is hopelessly limited. What is inside and what is outside the given organization is at best arbitrary. Our knowledge gaps in understanding this concept are so vast that little progress is made in designing conceptual schemes for exploring organization-environment interactions and interorganizational relationships through research.

11. Competition, Cooperation, Cooptation, Negotiation

Organizational systems are cultural answers to the problems encountered by human beings in achieving their collective ends. Power relationships develop necessarily from the different ways in which individuals can master the different sources of uncertainty relevant for their selective action[23].

Organizations use many strategies to accommodate to uncertainty in their environment. Katz and Kahn refer to these as control strategies:

Direct Control and Incorporation

The main organizational responses to uncertainty involve control, direct or indirect. Direct control can take two forms: internal regulation and external incorporation. Internal controls are aimed at producing a disciplined, unified system that can move quickly to meet environmental threats or changes.

Another form of direct control is aimed at the environment. If the environment can be controlled, it need not be predicted. Thus, an organization moves to incorporate critical parts of its milieu into itself. It extends its boundaries, for example, by purchasing the source of its raw materials, building a company town, or taking over competing firms. It attempts to influence clients and consumers, if not to control them directly, through expenditure on information and advertising.

Indirect Attempts at Control

Organizational attempts at direct control are often limited by the reactive character of an environment. It, too, has some organized character and the ability to fight back. The organization may therefore attempt to make itself less dependent on specific environmental conditions by building up some margin of reserve in its good years and diversifying its products or services in poor years. It may also seek to build support in the unorganized public and in organized groups through extensive public relations campaigns against some future crisis. In general, indirect attempts at control have replaced more direct forms as in the case of the developed nations and their industrial giants abandoning gunboat tactics for economic and technological measures.

Finally, organizations can reduce environmental uncertainty by gearing into larger systems, in order to take advantage of control mechanisms more powerful

149

than their own. Every organization falls back on the larger systems of societal values and legal norms to secure both general legitimacy for its activities and specific laws or edicts to buttress its position. Manufacturers try to keep foreign imports at a minimum by backing tariff laws. Corporations in the past have contributed to both major political parties in the United States, in the hope that they will be dealt with kindly by whichever party is in power.

Within a nation state, almost all organizations will take account of their environment by interaction with the political sector to assure legitimacy for themselves, to protect themselves against unfavorable legislation, or to gain economic advantage. Often these activities will be handled through secondary organizations such as professional associations, trade organizations, or organized lobbies.

Organizations thus use two strategies for overcoming the uncertainties of the changing world. The first we have described as direct control and environmental incorporation. An organization can expand beyond its original specialized function and seek to become a conglomerate, an educational institution, a political territoriality, or a welfare state. The second organizational strategy for controlling uncertainty is to maintain the boundaries around its main organizational function, and handle the many other tasks of political, informational, and supportive economic adjustment through interactions with outside organizations that specialize in these activities. As these interactions become stabilized, the organization in a sense becomes part of a larger social system.[24]

From *The Social Psychology of Organizations*, 2nd edition, by Daniel Katz and Robert L. Kahn p. 131. Published by John Wiley & Sons, Inc. Used with permission.

Katz and Kahn's control functions are similar to the boundary-spanning activities James Thompson described in the previous selection.

*Efforts at indirect control include influencing other systems by means of political manipulation or economic bargaining. In his book **Organizations in Action**[25] Thompson describes these strategies as efforts to arrange negotiating environments to acquire power. Strategies include contracting, coopting, and coalescing. The following excerpt defines these terms and partially answers the question of how an organization can maneuver towards achievement of power vis-a-vis its environment.*

Domains of Organized Action (an excerpt)

By James D. Thompson

Reprinted with permission from *Organizations in Action* by James D. Thompson, pp. 34–37. Copyright 1967, McGraw-Hill Book Company.

THE ACQUISITION OF POWER

Complex organizations "acquire" dependence when they establish domains, but the acquisition of power is not so easy. Organizations may, however, trade on the fact that other organizations in their task environments also have problems of domain and face constraints and contingencies. In the management of this interdependence, organizations employ cooperative strategies[26]. As Cyert and March[27] conclude, organizations avoid having to anticipate environmental action by arranging negotiated environments.

Cooperative Strategies

Using cooperation to gain power with respect to some element of the task environment, the organization must demonstrate its **capacity to reduce uncertainty** for that element, and **must make a commitment** to exchange that capacity.

Thus an agreement between A and B, specifying that A will supply and B will purchase, reduces uncertainty for both. A knows more about its output targets, and B knows more about its inputs. Likewise, the affiliation of a medical practitioner with a hospital reduces uncertainty for both. The medical practitioner has increased assurance that his patients will have bed and related facilities, and the hospital has increased assurance that its facilities will be used.

Convincing an environmental element of the organization's capacity to satisfy future needs is enhanced by historical evidence; prior satisfactory performance tends to suggest satisfactory performance in the future, and we might expect the organization to prefer to maintain an ongoing relationship rather than establish a new one for the same purpose.

Under cooperative strategies, the effective achievement of power rests on the exchange of commitments, the reduction of potential uncertainty for both parties. But commitments are obtained by giving commitments and uncertainty, reduced for the organization through its reduction of uncertainty for others. Commitment thus is a double-edged sword, and management of interdependence presents organizations with dilemmas. Contracting, coopting, and coalescing represent different degrees of cooperation and commitment, and present organizations with alternatives.

Contracting refers here to the negotiation of an agreement for the exchange of per-

151

formances in the future. Our usage is not restricted to those agreements which legal bodies would recognize. It includes agreements formally achieved between labor and industrial management via collective bargaining, but it also includes the understanding between a police department and minor criminals to forego prosecution in exchange for information about more important criminal activities. It also covers the understanding between a university and a donor involving, for example, the naming of buildings or the awarding of honorary degrees. Contractural agreements thus may rest on faith and the belief that the other will perform in order to maintain a reputation or prestige, or they may depend on institutional patterns whereby third parties can be depended upon to evaluate fulfillment of obligations and assess penalties for failure[28].

Coopting has been defined[29] as the process of absorbing new elements into the leadership or policy-determining structure of an organization as a means of averting threats to its stability or existence. Cooptation increases the certainty of future support by the organization coopted. The acceptance on the corporation's board of directors of representatives of financial institutions, for example, increases the likelihood of access to financial resources for the duration of the cooptive arrangement. But coopting is a more constraining form of cooperation than contracting, for to the extent that cooptation is effective it places an element of the environment in a position to raise questions and perhaps exert influence on other aspects of the organization.

Coalescing refers to a combination or joint venture with another organization or organizations in the environment. A coalition may be unstable, or may have a stated terminal point; but to the extent that it is operative, the organizations involved act as one with respect to certain operational goals.

Coalition not only provides a basis for exchange but also requires a commitment to future joint decision making. It is therefore a more constraining form of cooperation than coopting.

DEFENSE OF DOMAIN

The attainment of a viable domain is, in essence, a political problem. It requires finding and holding a position which can be recognized by all of the necessary "sovereign" organizations as more worthwhile than available alternatives. It requires establishing a position in which diverse organizations in diverse situations find overlapping interests. The management of interorganizational relations is just as political as the management of a political party or of international relationships. It can also be just as dynamic, as environments change and propel some elements out of and new elements into a task environment.

And just as political parties and world powers move toward their objectives through compromise, complex purposive organizations find compromise inevitable. The problem is to find the optimum point between the realities of interdependence with the environment and the norms of rationality.

The public school, for example, which is constrained to accept virtually all students of a specified age, under conditions of population growth has urgent need for power with respect to those in the task environment who control financial and other inputs. If the task environment imposes mandatory loads, the school must seek power with respect to resources. The private school, on the other hand, may be able to read both student load and inputs as variables, and seek their mutual adjustment.

The business firm constrained by an impoverished market, as during a recession, finds it urgent to have power to curtail the rate and price of inputs provided by supply elements of the task environment. To the extent that it has power, it may renegotiate contractual arrangements. If the firm is also constrained by large fixed costs, as in heavy industries, our proposition would predict that the organization will seek power to curtail the flow of labor inputs. It is in such industries that wage payments typically are in hourly or piece rates, and firms are not committed to fixed salaries or guaranteed annual wages. By contrast, in the university, where variations in student load occur primarily at only one time of the year, wage payments are in annual terms. The organization facing many constraints and unable to achieve power in other sectors of its task environment will seek to enlarge the task environment.

Captive organizations frequently find themselves boxed in on several sides, to the point where norms of rationality are threatened or overwhelmed. It is at this point that captive organizations often join forces to establish noncaptive evaluating organizations which develop yardsticks of rationality and set standards for accreditation. Community hospitals, prisons, city governments, and public schools all exhibit this device of creating new elements in the task environment to offset other constraints within it. To the extent that the new element has power to confer or withhold prestige, it can loosen the constraints operating on the organization. The nonaccredited school or hospital, for example, may be threatened with irreplaceable loss of personnel, to the point where those who control financial inputs are forced to increase their support.

Nursing Aide Training by Community Saves Hospital Time, Space, Money

By Harry V. Berg

Reprinted, with permission, from *Hospitals, Journal of the American Hospital Association*, volume 52, Number 18, September 16, 1978, pp. 91-96.

In the next selection, Harry V. Berg describes how a hospital gained power over its supply of well-prepared nursing aides through cooperation with another hospital. The administration looked beyond the walls of its institution for the answer to its problem of diminished resources. Note the by-products of this cooperation strategy. The hospital released the time of RN staff, which was then available for patient care; second, the exchange with the other hospital set up a communication pattern which has the potential for facilitating similar negotiations in the future.

In mid-1975, the staff development department at W. A. Foote Memorial Hospital, Jackson, MI, was faced with an increasing number and variety of assignments and was forced to find means to increase its productivity without adding personnel or increasing its budget in any way. The department decided to determine which of its assignments or tasks required the greatest resources, and it was almost immediately apparent that the resources required for nursing aide and orderly training exceeded those required for any other activity.

Nursing aides and orderlies provide a very important part of the hospital's patient care services, and their training is vital. However, because this training required so much of the total staff development effort, the department very carefully examined the direct labor costs, the space, and the other requirements of the nursing aide and orderly training classes to determine whether any reductions could be made.

Approximately $40,000 per year in wages was being paid to the RN instructors and approximately 20 trainees in each of the four classes conducted each year. Books and other instructional materials also contributed to the costs. The training classes required 16 weeks of time each year from two of the RNs in the staff development department, thereby drastically reducing the time available to them for numerous other important educational priorities. In addition, securing space in which to conduct the program was a continuing problem, and the

155

available space was barely adequate. Thus, achieving savings in the money, space, and time that were committed to nursing aide and orderly training was an urgent goal for the staff development department.

Because the educational training priorities of the hospital's nursing department were involved in this issue, it was necessary for the staff development department to discuss its needs and ideas with the director of nursing. She was very supportive of the department's ideas. Her chief concern was that the instructional quality of any new training approach that would be adopted would maintain the skill levels required of the hospital's nursing aides and orderlies. The department promised that continuing high standards would be ensured and that, otherwise, it would not make the changes.

OUT-OF-HOSPITAL PLAN

The department decided to have an outside community agency conduct the nursing aide and orderly training. First, it secured the approval of the personnel director for this plan, and then it determined which outside agency would be interested in and competent to conduct the training. The adult education division of the public school system seemed to be a good prospect, and it in fact turned out to be a very good choice. The division's only reservation about the plan was concern about the design of such a specialized curriculum, but this concern was relieved when the department agreed to turn over to the division its entire curriculum and its teaching methods. Thus, the outside training program is the same as the former in-hospital program. The department also agreed to provide on-the-job hospital experience for the students, under the supervision of the two RN instructors.

Before planning could be completed, a school funding problem threatened to halt the whole project for an indefinite time. Consequently, the staff development department and the adult education division turned to two federal government agencies for support—the WIN (Work Incentive Programs) and CETA (Comprehensive Employment Training Act) agencies. These hard-working federal and state funding agencies provide excellent service in funding educational programs that help unemployed persons get jobs. Their ability and willingness to help saved the training program.

The hospital and the adult education division then jointly established an adult nursing aide training advisory committee, with the public school system's adult education director as its chairman. Representatives from the following organizations attended the initial meeting, and most of them became regular members of the advisory committee: W. A. Foote Memorial Hospital; adult education division, Jackson public school system; Jackson County Medical Care Facility; the Jackson WIN agency; the local CETA agency; Jackson County's intermediate school district; the local department of social services; Region II manpower consortium; and Jackson Osteopathic Hospital.

Until graduation of the first class of nursing aides, this group and its subcommittees met numerous times. For example, the curriculum subcommittee was concerned with applying the hospital's educational material to the public school system's adult education program. The committee also dealt with issues such as quality of instruction, student/teacher ratio, total number of hours of instruction, methods for screening prospective participants, hiring instructors, selecting textbooks, methods for evaluating students' progress, and related issues. The participating parties agreed to structure the program as follows.

Funding for the program would be provided by the local WIN and CETA agencies. This funding would pay the salaries of the two RN instructors and for uniforms, books, supplies, travel, and other costs for the welfare trainees in the program. Most of the students would be welfare trainees, but provisions were made to enroll some trainees who would not need WIN or CETA help.

The WIN staff would interview and screen all applicants to determine whether they meet the requirements for entrance into the program.

The public school system's adult education division would have overall control of the program—for example, to hire and pay the teachers, to maintain class records, and so forth.

The space in which to conduct the program would be provided at no cost by Jackson County Medical Care Facility as its contribution to this communitywide cooperative effort.

To begin the program, the public school system's adult education division hired two well-qualified instructors. In this initial effort, the hospital played a major consulting role for interviewing and hiring the two RNs. Since that time, the public school system has done an excellent job of hiring and paying instructors and handling all other details of the course.

Good discipline has been maintained in the program. For example, after one absence from class, a student is provided with a counseling session, and after two absences, he is dropped from the program. The concentrated learning pace of the condensed training program makes strict attendance essential, because catching up on lost time and missed work has proved to be extremely difficult for most of the participants. Thus, the discipline is protective rather than punitive. If a student is dropped from the program because of absences that are attributable to illness, he is afforded top priority for admittance to the next class.

RESULTS, GOALS ACHIEVED

In the first year the program was in operation, five classes for nursing aides and orderlies were conducted. Of the 102 students who began the course, 82 earned certificates of achievement. Of these 82, 77 were employed, primarily at either W. A. Foote Memorial Hospital or Jackson County Medical Care Facility, and two continued their studies at Jackson Community College in order to become LPNs. Thus 93 percent of the graduates either were employed or were continuing their education.

Graduates from this program gain status as thoroughly trained and competent personnel. Evidence of the program's continuing success is the waiting list of prospective students. The certificates of achievement are well recognized in the community, and the training generally leads to worthwhile jobs. Consequently, the participants have great respect and appreciation for the program.

The program achieved the goals of the hospital's staff development department.

First, the time that the department devoted to nursing aide and orderly education was reduced from 80 days of training per year to 12 days of orientation per year, thereby saving 68 days per year that can be used to meet other high-priority training and educational needs.

Second, the outside program reduced the wages paid for instruction by approximately 85 percent, and it reduced the wages paid to trainees to those for the orientation program[30]. The net savings have been approximately $36,000 per year.

Third, the need to find space for the classes in the hospital was eliminated, because Jackson County Medical Care

Facility was able and willing to provide the space.

In addition, the other patient care institutions in the community were provided with a pool of trained nursing aides for potential employment.

Our readings would lead us to conclude that the particular strategy an organization uses to achieve its ends in the face of environmental uncertainty is highly dependent upon its structural characteristics, its bargaining position, its capacity to satisfy the input-output needs of its task environment, the extent to which it is able to monopolize this capacity, and the availability of scarce and valued resources.

For example, in an urban Midwest community several years ago, 19 hospitals were in fierce competition for professional nursing skills, which were in short supply. New nursing graduates would freely shop around for the most attractive financial offer in a seller's market, causing each hospital constantly to try to outdo the others' starting salaries. Realizing that they were losing control of the situation, the hospital administrators, through their council, banded together and cooperatively agreed upon a starting salary. This was effective for a short time, but then several hospitals began to use retroactive tuition reimbursement as additional negotiating leverage. The hospitals without schools of nursing were not able to use this additional employment incentive legitimately. They were at a distinct disadvantage until the supply of new graduates in the community became more adequate.

In this example, the lack of assurance of a sustained input (nurses with appropriate skills) to maintain the services of the hospital led to organizational response to reduce uncertainty. Though the hospitals normally compete for patients, they set aside their normal competitive practices to gain control of their input by coalescing. The administrators used their professional association as the controlling mechanism for an interorganizational network. Though there was temptation to violate the joint agreement, care was exercised to choose a method which had a semblance of legitimacy, that is, tuition reimbursement.

William Evans describes such strategies as follows:

Organizations capable of modifying their environments engage in what is essentially political action. They use their resources and sanctioning power against actual or potential opponents to alter events—role behavior of target individuals, organizational behavior, law, and other institutionalized arrangements.[31]

Political action is used here in the sociological sense: the control and management of people living together in a society. As mentioned in the first unit of this anthology, the political connotations of interorganizational networks have led to increased study of the political realities which affect organizational structure and functioning. An example of research into the characteristics of interorganizational politics is that of Kenneth Benson.

Kenneth J. Benson[32] has synthesized the research of many scholars to find commonalities which can form a general framework for understanding the special political-economic problems of human service interorganizational networks. His formulation concentrates on the distribution of two scarce resources, money and authority. Authority refers to the legitimation of activities, the right and responsibility to carry out programs of a certain kind. Benson maintains that the interactions and sentiments of organizations, which are bound together in a network, are dependent upon their respective market positions and power to affect the flow of resources. The interorganizational network itself is linked to a larger environment consisting of authorities, legislative bodies, bureaus, and publics. The flow of resources into the network depends upon developments in this larger environment. By this time, you will recognize the similarity between Benson's theoretical conceptualization and Yuchtman's. Benson talks of market position; Yuchtman uses the term bargaining position. Benson refers to power to affect the flow of resources; Yuchtman calls it the ability to acquire scarce resources. The primary difference between the two conceptualizations is that Benson is using an interagency network as the unit of analysis, whereas Yuchtman used one organizational entity as the unit of analysis.

Benson identified four general strategies for organizations to use to change their network relations. These are:

1) cooperative strategies in which change is sought through agreements and joint planning, 2) disruptive strategies in which resource-generating capacities of agencies are threatened, 3) manipulative strategies in which the supply of and sources of the resources money and authority are tactically altered, and 4) authoritative strategies in which network relations are precisely fixed by prescriptive action of resource controlling agencies, offices, or bodies.

Benson classified organizational sentiments and interactions into four analytical dimension. These are:

1) domain consensus or agreement between agencies on the role and scope of each, 2) ideological consensus, agreement between agencies on the appropriate approaches to their common or similar tasks; 3) evaluation or the judgments between agencies of the quality of work of each; and 4) work coordination, referring to the conduct of joint, cooperative, or articulated activities and programs. These four dimensions form a system of interrelated variables which vary together. The system tends toward balance in the sense that high levels on one dimension tend to be associated with high levels on others. A strain toward balance is assumed. Imbalance of the system is said to create a tension or strain toward consistency.[34]

Interorganizational power, whatever its source, may have a variety of effects upon network relations. Benson asserts that administrators in powerful agencies should be able to defend their resource flow, claim new domains, resist claims advanced by less powerful agencies, and block the creation of competitive organizations. Power permits one organization to reach across agency boundaries and determine policies or practice in weaker organizations.

Whole professions use this kind of political power. A case in point was the attempt by the American Medical Association to affect the role of the nurse through public statement at its annual convention in 1970.

For an outstanding discussion of the role the medical profession plays in the political economy of health care delivery, E. Freidson's **Professional Dominance** *is highly recommended.*

The next selection is a study of how community pressure groups (including physicians) affected the ability of rural community hospitals to function. Many of these dynamics occur with urban hospitals as well, though taking a somewhat different form—today Health Systems Agencies (HSA) have replaced the Regional Medical Programs and the Comprehensive Health Planning Agencies. This type of social action research makes enjoyable reading. While it lacks the rigor of the more arduous studies we have been examining, it is no less useful in alerting us to environmental factors which the nurse administrator should consider when planning organizational change.

STATIC DYNAMICS IN MEDICAL CARE ORGANIZATION[36]
By Stanley R. Ingman

Reprinted with permission from *Topias & Utopias in Health,* edited by S. Ingman and E. Thomas. Published by Mouton Publishers, 1975.

The voluntary community hospital remains the dominant institution in the medical care arena with respect to controlling the organization of medical care manpower and resources. Physicians still dominate hospital policy formulation as well as community health policy. This situation prevails in most rural areas in the United States and must be analytically examined before medical care reorganization can be proposed.

I shall depict in broad terms how the multiplicity of social structural preconditions and societal ideologies determined medical care in one rural community hospital over a twenty-year period. Specifically, I shall attempt to describe both the external influences upon this rural community hospital as well as the internal dynamics that jointly determined the quality and quantity of medical care offered within this hospital. In addition, the health planning phenomenon of the 1960s will be examined and evaluated with respect to why it has had so small an impact upon the quality and quantity of rural medical care.

Among the formal agencies and groups that influenced the delivery of hospital services in this community, the most significant were: the county Medical Society, the Joint Commission on Accreditation of Hospitals, the State Hospital Board, the Regional Medical Program, the Comprehensive Health Planning "B-Agency," the State and National Medical Associations, and the School of Medicine. Informal pressure groups in the health arena, such as labor unions, corporations, and voluntary health and welfare agencies within and outside the region, roughly complete the social network of influences within which this community hospital evolved. The protagonists within this community hospital included the trustees, the physicians on the medical staff, and the hospital administrators whom the board had appointed.

Although this analysis is of one rural community hospital, a quick review of nearby hospitals suggests that it is not markedly atypical in terms of quality and basic organizational structure and purpose, i.e., division of labor, resource allocation, decision-making process, and legitimating goals and objectives.

The data were secured in part through careful review of files that had been collected for over a twenty-year period. Interviews with various health professionals, administrators, and community leaders served to clarify and expand upon the materials in the files. The local newspaper archives helped provide additional verification.

Essentially, it is a basic contention of this work that local leadership—including both health providers and lay persons—is unable to create the quality and quantity of health services that might be reasonably expected in such a region in the 1970's. A corollary is that external influences, such as those previously mentioned, are inadequate, either singly or in combination, for protecting the general welfare in rural community hospitals. I will try to provide an illustration to support these contentions[37].

"The Dockston" Hospital[38] is a 200-bed, general acute-care hospital under voluntary sponsorship located in a semirural, economically depressed river town twenty miles north of the large urban center of "Metroville" within the Appalachian region. The hospital is characterized by a slightly atypical board of directors. For example, two members are politicians in the next town to the north. Otherwise, with few exceptions the board of directors represents the reputed leaders in the valley rather than formal political representatives.

In 1949, the board hired a young administrator who had no actual experience or formal training in hospital administration. However, this young man diligently pursued his work, read widely in the area of hospital administration, and soon became more knowledgeable about modern hospital patient care. He continually introduced innovations into the administrative structure of the hospital. As a representative of the new breed of what is called "clinical administrator," he also began to feel more and more that the medical care offered in his hospital ranked below the standards of a modern hospital. Consequently, he tried to encourage the physicians to reorganize themselves in order to improve care. The hospital

surveys conducted by the American College of Surgeons in 1945 and 1948 and later by the Joint Commission on Accreditation of Hospitals in 1955, 1958, 1961, 1964, 1965, and 1968 continually gave credibility to his prodding of both the boards and the staff to act. His strategy of persuasion won some minor success but basic reorganization was not accomplished during his stay at the hospital, which lasted eighteen years.

The medical staff must be given the major share of the responsibility for the medical care provided at the "Dockston" Hospital. The physical plant and administrative operation only indirectly affected the level of care offered at this hospital. On paper, the board held the official authority; yet I contend that in most rural hospitals in the United States this authority is typically confined to narrow limits for reasons that will be discussed (see[39]).

At the "Dockston" Hospital the staff was controlled by an older group of general practitioners who over the years had become accustomed to having their own way without any interference from the board or the administrator. These general practitioners wanted the institution to remain limited in scope. They did not think it was desirable for the hospital to attract new specialists, nor was it thought feasible for the institution to enlarge its staff markedly. There was a small group of certified specialists on the staff who mostly disagreed with the general practitioners. Both the younger general practitioners and specialists who were of this persuasion found it hard to become involved in the policy making of the medical staff. While national trends have been on the side of specialists, the general practitioners held the power in the "Dockston" Hospital.

Because there were no board-certified surgeons and gynecologists on the staff, general practitioners performed most of the work in surgery and gynecology in the hospital. In the 1950's there were no specific departments and, thus, most older general practitioners peformed whatever tasks they felt qualified to perform.

In the late 1950's and early 1960's the older general practitioners began to resent the persistent agitation for change, especially by the administrator. In order to form a united front in their effort to obtain accreditation in 1958, the administrator was taken into the confidence of the staff. Later, a few physicians felt that the administrator was using this inside information to "brainwash" the board against them. The medical staff leadership also noticed that a small group of the younger staff members was more militant in their thrust for more prerogatives.

In late 1962 the old guard began to feel that the opposition had increased too rapidly over the preceding years, and they attempted to reestablish their authority once more. They focused their criticisms on the administrator, whom they envisioned as the leader of the "conspiracy." The staff felt that the board and the dissident members of the staff would lose their courage if and when the administrator were eliminated from the hospital. Without calling for a staff vote on the matter, a few officers went to the board and demanded that it fire the administrator, claiming that the administrator was meddling in their affairs and in general was very annoying. The board was somewhat shocked at this demand and indignantly refused to honor the staff's request. Instead a consultant was secured, and a major public dispute soon erupted into the courts and the public media.

THE HOSPITAL: INADEQUACIES AND RECOMMENDATIONS

At the crux of the dilemna facing the hospital administrator, the staff, and the board was the quality of health care offered at the "Dockston" Hospital. The staff had been criticized as early as 1945, when the American College of Surgeons "seriously considered withdrawing their approval of the "Dockston" Hospital because the staff was not meeting the minimum requirements for hospital standardization"[40]. It was indicated that some problematic issues were not new to the board, e.g., "the poor condition of the medical records." It was also suggested by the associate director of the American College of Surgeons that monthly medical staff conferences with audits of clinical work were not being conducted, although these were required for accreditation.

The following needed changes were discussed: more consultations, a clinical laboratory facility, more autopsies, and the establishment of a department of surgery. The secretary of the board noted in 1945 that "a medical audit may from time to time deflate the ego of some of the staff, but if it saves a single life, is it not worth it?"[41]

In 1948, a surveyor arrived to determine whether the hospital deserved continued accreditation. The hospital scored 72 percent, 3 percent above the limit of the provisional approval level (60-69 percent). The hospital lost points primarily in relation to seven major areas: medical staff organization, medical record department, clinical laboratory, medical department, surgical department, obstetrical department, and medical social service department. Other comments dealt with the overcrowded facilities, the failure of some members to sign medical staff by-laws, the fact that there were no regularly appointed consultants, no regular presentations and discussions of hospital cases, no special clinico-pathological conferences, little improvement in medical records, an autopsy rate far below the minimum (2/185 in 1948), a need for more nursing supervisors, a general lack of qualified doctors[42] (fellows or diplomates) in the medical, surgical, and obstetrical departments, a high rate of infant mortality resulting from Caesarean sections, and, finally, the fact that there was no outpatient department. Full approval was given on the condition that steps be taken to correct the above-mentioned deficiencies.

It was not until 1955 that the hospital was again reviewed by an outsider. This time it was the Joint Commission on Accreditation of Hospitals (JCAH), which had been formed in 1952. There were sixteen different deficiencies outlined, and accreditation was denied. Many areas were labeled inadequate, e.g., medical staff committee meetings and minutes, the autopsy rate, medical records, the consultation rate, and the Caesarean section rate (twice the national average). Included in the review was the suggestion that they revise the staff by-laws and thereby its organization and that they appoint a joint conference committee in order to establish a liaison between staff and board (this last suggestion was quickly adopted).

After three years the board and the staff agreed to receive a new surveyor from the JCAH. Neat black books were prepared and a specific agenda was planned to "coordinate" the surveyor's visit. It proved to be a successful strategy and accreditation was granted. Some felt that accreditation was gained as a result of the neat black books rather than improved standards. The fact that the surveyor was a general practitioner was also thought to explain accreditation in 1958.

However, in 1961 the hospital attained a full three-year accreditation. It was again noted that the rate of autopsies should increase, obstetrical and medical records should contain more clinical information, and operative permission and procedures should be reviewed.

The normal JCAH survey cycle of every three years was interrupted in 1962 by a dispute between the staff and the board over the staff's desire to have the administrator fired. As a result, the board decided to hire a hospital consultant to settle the dispute and to restore harmony within the hospital. The executive director of the hospital council of the eastern part of the state recommended a nationally noted consultant, "Dr. Fargue." In November 1962, he and his associate began their survey.

In December 1962 the report was completed but was distributed first to only two board members on the executive committee, an attorney and a retired "Metroville" steel company executive. Later two other members were given copies to read. The report covered hospital utilization in "Coal Valley," board organization, hospital administration, and medical staff performance. The report was never made fully public, or even circulated among all board members.

According to the surveyors, hospital utilization was low (80 percent). Since 1955, the United Mine Workers (UMW) Welfare and Retirement Fund had eliminated the hospital from its approved list, and it was suggested that a new agreement be arranged between the hospital and the fund. A major stumbling block to negotiation was the fact that the hospital would not permit physicians in the "Circleville" UMW Medical Group to be on the staff. Before any agreement could be reached, this rule would have to be changed.

All three hospitals in the area showed a low occupancy rate in the maternity service. To achieve regional economic rationality it was suggested that some consolidation of the three maternity services would be advisable. It was also suggested that the hiring of a specialist in obstetrics would represent one way of increasing the occupancy of the maternity service. It would have required opening discussions with the two other hospitals in the area and the "Circleville" Medical Group.

Although the administration of the hospital received an excellent rating, the board was given some suggestions. First, the board was asked to modify its viewpoint with respect to "its rights and duties." A book on hospital trusteeship was handed to each board member to help transform his views. The report pointed out the need for extensive changes in the constitution and by-laws of the board, e.g., the consultant felt that the board had apparently delegated inapproprately certain "authority" to the staff. The surveyor rejected the claims voiced by some physicians that the administrator was incompetent and detrimental to the operation of the hospital. It was recommended that the board continue to support him.

According to the review, the staff organization looked good on paper, but did not seem to function according to the standards of hospital accreditation. It was recommended that new medical staff by-laws be adopted. It was also suggested that several new committees be added (a credentials committee, a medical audit committee, and others).

There were numerous charges of inadequacies. Surgery privileges were thought to be uncontrolled or left to the individual physician's discretion. Partly because "qualified" physicians and surgeons were lacking, complex cases in the area were often sent to "Metroville" hospitals. However, patients were referred to "Metroville" for other reasons as well. For example, although some physicians stated that local men were incompetent, others contended that they would be more assured of not having patients stolen by local colleagues if they sent patients to "Metroville" for specialty care. It was thought commendable for the hospital to refer complicated cases to "Metroville" but it was noted that such a procedure did

result in a lower occupancy rate for the hospital as well as limiting the scope of the hospital.

The consultant suggested that the number of specialists be increased so that the scope of services at the institution could be expanded. It was also mentioned that although the hospital had adequate departments of radiology and pathology, they were functioning below 50 percent of their capacity. This could be explained partly by the staff's limited use of diagnostic and review procedures. To satisfy the need for physicians who were certified in obstetrics and gynecology, it was suggested that the specialists practicing at the "Circleville" Medical Group and at the "Blacksburg" Hospital be invited to head up departments at the "Dockston" Hospital. The consultant summarized his remarks about the staff in the following manner: "The physicians on the staff of the 'Dockston' Hospital are good family doctors and are practicing a brand of medicine which is adequate for family doctors but is not adequate for a modern hospital or medical center . . . "

The consultant also illustrated the staff's inadequacies in his review of their committee work. The staff meeting had good attendance but failed to review the quality of medical work performed in the hospital. It was suspected that the tissue committee, although producing excellent minutes, never met, as was also the case with the infectious diseases committee. Therefore, the pathologist was never given the opportunity to attend the tissue committee meetings as required by the JCAH. It was indicated that the surveyor from this commission had been told otherwise. The low autopsy rate was again cited, as well as the slipshod means by which consent for operations was secured. The reviewer reported that the medical records committee and members of the department of surgery were holding meetings. In summary, he felt there was an overall need to revise the staff's by-laws, rules, and regulations.

The medical audit raised many serious questions about the quality of work performed at the hospital. As cited by the reviewer, the major problems lay in the inadequate review of clinical work done in the hospital, e.g., record keeping was incomplete for obstetrics and newborn cases, and review was lacking in cases of death. Preparatory tests and consent forms were thought to be underutilized, e.g., few pregnancy tests were recorded, and consent forms and consultations were lacking in some cases of hysterectomy, sterilization, and other surgical interventions. The medical audit, in general, showed a lack of supervision, control, review, and analysis of the surgery done in the hospital. The report did not claim that any generalized abuses of surgical privileges were occurring, but suggested that departmentalization might encourage better internal supervision.

In October and November of 1963, a special committee of the staff invited four prominent physicians from a nearby school of medicine to advise the staff on reorganization. Professors in internal medicine, surgery, pediatrics, and obstetrics and gynecology came (each on a separate occasion) to visit the medical staff. In brief, although all of the visitors were critical of the big city consultant for his coercive style, of the administrator for "practicing medicine," and of the board members for being unduly influenced, they emphasized the need for obtaining more specialists on the staff, for instituting limited departmentalization, for hiring a medical director, for instituting open staff policy, for utilizing more consultants, and for revising the medical staff by-laws. The professors based their recommendations on one afternoon's discussion with the leadership of the medical staff.

In the fall of 1963, the county medical staff became formally involved as a result of pressure from the American Medical Association and the state medical society. The chairman of the county medical society review committee wrote the hospital administrator: "considerable pressure has been brought to bear upon the committee . . . to render a report . . . we have heard complaints of the Medical Staff . . . we want to meet the Board. This matter has caused great concern to organized medicine, both at the State and National level."

After its visit with the board, the committee made its recommendations. The report began by proposing that the staff should expand, improve, and supervise their own peers more closely; it also stated that the board had been justified in its actions. The medical society's committee felt that the choice of the consultant had been unfortunate and that the consultant had engaged in "dictatorial, abusive, and unjustified criticism." The real problem, they believed, was the loss of confidence, respect, and reasonableness on both sides. They called the board's action "revolutionary" and the staff's reaction "abusive and vindictive." They supported quality control and the introduction of more specialists, but they instructed the administrator to cease interfering in medical affairs. They also suggested that a medical director by appointed. Although they agreed that the staff by-laws needed revision, they suggested that the control of medical practice in the hospital should be returned to the staff. They felt that the key to the problem was the maintenance of proper communication channels.

In that same year the "big city" consultant returned to review the hospital. Although he noted certain improvements, he pointed out some of the same problems he had discussed earlier, e.g., the low autopsy rate, the board's failure to exert its legal authority ("have decided to settle for a mediocre hospital''), the failure to control surgical privileges, and the unresolved feud with the UMW clinic.

In 1964, the JCAH sent a specialist to review the hospital. The specialist granted accreditation for only one year. Twenty-four specific recommendations and comments were made in the survey, e.g., the tension between staff and administration was dysfunctional, there was a need for a written policy on the use of oxytocic drugs, there was a need for departmentalization, the autopsy rate should be increased, and surgery privileges should be checked. However, a general practitioner from the JCAH resurveyed the hospital in 1965 and gave the hospital a three-year accreditation. Accreditation was renewed for three more years in 1968.

INTERNAL DYNAMICS

In this section I shall present an abbreviated discussion of the social conflict between the board, the physicians, and the administrator. The genesis of the conflict can be traced back to the period when the administrator, who had become aware of the inadequacies at the hospital, attempted to indicate to the board its responsibility for improving the hospital. Although a few young specialists supported this new activist orientation, the medical staff leadership resisted passively at first and then later set out to destroy the reform movement by practically every means available to them.

However, both the reformers and the medical staff leaders became equally upset when the hospital board moved to appoint medical specialists from the nearby United Mine Workers clinic. The resistance from the hospital reformers at the hospital indicates that there were specific limits to the reform plans. They were unwilling to have the specialists from the so-called "socialist"

UMW clinic help fill the specialist gap that had been pointed out by all the hospital reviewers. Therefore, although the board met with the UMW physician representatives to negotiate possible appointments, the resistance from the entire membership of the medical staff quickly discouraged any further negotiations.

Several months after the consultant's medical audit was completed, one staff member went so far as to steal the report from the administrator's office. When the board threatened legal action, the report was returned. Following this incident, other physicians covertly helped organize a citizen's group to apply pressure on the board to stop their plans to reorganize the medical staff. The theme of the group was that the board was going to run all of the physicians out of town. The medical staff leadership also sponsored advertisements in the local paper and some physicians attacked the board and the administrator on the local radio. As the board moved ahead with its plans to rewrite the hospital and medical staff by-laws, the staff secured two separate court injunctions to stop implementation. After recalling that the judge actually cross-examined the board members and openly supported the medical staff leadership, the rumor that the judge had been bribed seemed plausible.

Other pressures began to be applied. Prominent local entrepreneurs on the board were threatened with loss of business. With the court action and with apprehension about economic reprisals from both angry citizens and physicians, the board's activism dissipated markedly. Managers of absentee corporations in "Coal Valley" (some of the strongest reformers on the board) began to question the potential consequences of alienating local citizens in the name of improving the quality of medical care. Their reluctance to candidly tell the public "the whole story" was based on their fear of completely destroying public confidence in the hospital and their leadership. This fear of mass reaction kept citizens ill-informed throughout the ten or more years of the dispute.

The directors and owners of these absentee corporations began to discourage their local managerial representatives from remaining "responsible" and "involved," especially when it appeared that the hospital controversy might backlash[43]. They feared that their own corporate operations in the valley might receive more careful attention by the citizens; for example, low corporate taxes might be questioned. With all supports backing down, the hospital atmosphere became so stressful for the administrator that he suffered two heart attacks and left the hospital. Although he vowed never to become involved in medical care administration again, two years later, a UMW clinic outside the immediate area managed to recruit him as their administrator.

The few physicians who had supported the board either left the area or accepted a partnership with one of the local physicians in the leadership group. Thus, the coalition between some board members, some specialists, and the administrator collapsed. The group legally responsible for ensuring the public welfare, namely the board, was unable to maintain its push for change.

EXTERNAL INFLUENCES

Hospital Accreditation

In the United States, where voluntarism reigns, medical and hospital professionals are very proud that they established the American College of Surgeons (1913) with a hospital review system, and later the Joint Commission on Accreditation of Hospitals. The first systematic review of hospitals took place in the United States in 1918 and 1919. Only eighty-nine out of 671 hospitals were approved as meeting the Minimum Standards for Hospitals drawn up in 1918. However, the list of hospitals approved and those disapproved was destroyed because too many important hospitals failed to pass inspection. The professional hierarchy worried about the terrific repercussions if this information were to be made public.

In 1952, the JCAH was formed from the American College of Surgeons (three votes), the American College of Physicians (three votes), the American Medical Association (seven votes), the American Hospital Association (seven votes)[44]. Although the new hospital reviewing board broadened the representation, it is essentially still a medically dominated body. However, it is the case that the American College of Surgeons and the American College of Physicians agree more often than not with the American Hospital Association.

In 1964, accreditation was almost denied to the "Dockston" Hospital, but after some pleading by the administrator, who warned the surveyor that the valley would explode if accreditation were to be denied, the JCAH granted a one-year accreditation. In 1965, a new surveyor essentially presented them with a "clean bill of health" and accreditation was granted for another three years. This was repeated in 1968 as well.

The past history of the "Dockston" Hospital was easily accessible to the new surveyor. The JCAH office, which makes the final decisions, was well informed about the history of the hospital, e.g., two staff members (from the leadership group) had visited the Chicago office in an attempt to explain the staff's side of the argument and had asked the JCAH to investigate the hospital. Although there were some observable changes in the character of the staff performance, it is hard to correlate improvement with accreditation at this hospital. In this instance, the JCAH seems to have avoided active involvement because of the controversy.

As a catalyst for change, the Joint Commission can be given some credit, but the voluntaristic ideology by which it is explicitly guided seemed to prevent it from serving as an effective advocate for hospital standards. The view of "let the locals decide their own affairs" or "we (JCAH) are not set up to punish but only to encourage" releases the commission from any responsibility for the medical care in the "Coal Valley." But if this is true, then who has the obligation to look out for the public welfare?

The incongruity between different surveyors' evaluations and the level of actual standards at the hospital raises serious doubts about the meaning of hospital accreditation as well as the efficacy of the Joint Commission to guard the public interest. Having professional colleagues judging and evaluating each other definitely has its limits with respect to public accountability.

Big City Consultant

Perhaps the consultant underestimated the potential of the staff hierarchy to resist reform. He did not correctly assess either the level or the types of value commitments of the staff or the board members. The extent of, or lack of, vested interests of all parties in the dispute was not evaluated. It is true that the consultant's confidence, knowledge, and convincing presentation as the expert incited the board members initially to act aggressively with little thought about staff reaction. Their boldness reflected a naive understanding of the emotional and ideological commitments of the staff as well as the economic interests that were threatened by their reform program. The consultant failed to prepare the board for the ensuing

battle, which he should have been able to predict. If a consultant is truly concerned with having his suggestions adopted, he should devote considerable effort to informing the board of a hospital as to the possible consequences of their proposed reforms in realistic terms.

The majority of the staff was not motivated solely by economic considerations. Many expressed deep concern with their image as trustworthy physicians. Because of the lack of careful communication between each staff member and the board, some staff members became increasingly supportive of the status quo and violently upset with the board. Some physicians even retired. The circulation of false information by some staff members helped stimulate the wrath of men who otherwise might have remained cooperative with the board. However, because of the extreme personal attacks that were continuously inflicted upon the consultant and the board, their desire to find a harmonious resolution of the conflict dissipated rapidly.

Courts

In the courtroom in "Branburry" the "big city consultant" was portrayed to the public by "Judge Parks" and by the staff hierarchy as an outsider attempting to interfere in local affairs. Under this public characterization, the consultant personified to the public the big city expert who had "brainwashed" the board to follow his every wish. The judge, especially in the second hearing, also attempted to discredit any physician who might agree with the board or with the consultant. For example, in the courtroom the judge asked whether one physician, who had agreed to the new by-laws, was related to the administrator. (In fact, he was the administrator's brother-in-law.) In another instance, he attacked the attorney of the board for conflict of interest because he was also a member of the board. By leading witnesses to divulge particular facts, he attempted to challenge the board's credibility.

The question must be raised as to why he behaved in this manner. Was the judge honestly hostile to an outsider pushing his fellow professionals and friends around? Was the judge truly antagonistic to a board of laymen trying to regulate and control physicians who were fellow professionals? Although there may be some reasons to reply in the affirmative, there exist other underlying facts that explain his overt support of the medical staff in the courtroom.

Several informants claimed that "Judge Parks" received $8,000 from several physicians on the staff for his support. One board member contended that certain physicians had such political influence in the county courthouse that it was quite unnecessary for money to change hands. Another informant claimed that two physicians on the hospital staff had boasted to him that they controlled "Judge Parks."

Nevertheless, it is evident that "Judge Parks" went out of his way to discredit the board and its allies. Recalling "Judge Parks' " connection with racketeers in the 1950s, this was not the first time that this judge had been rewarded for his cooperation. Naively, the attorney for the board, knowing of similar cases in which courts had ruled in favor of a hospital board, felt confident that the court would rule in its favor. He seemed unaware of what was awaiting him and the board in the county courthouse.

Organized Medicine

As the dispute began to attract public attention, the American Medical Association became aware of the ongoing feud and notified the State Medical Society to have it resolved. The officer of the state society in the "Metroville" area contacted the head of the County Medical Society. In December 1963, the county group contacted the hospital board, offering its help in resolving the dispute.

The Committee for the County Medical Society met separately with the medical staff and the board. Subsequently the committee presented its final report to the board. The committee agreed that the board was justified to act, but stressed that the consultant had been abusive and that the administrator should stop "practicing medicine." The members also stated that the staff had behaved poorly and had engaged in campaigns of personal vilification. Poor communication and misunderstanding were the major problems to be rectified. The committee essentially agreed with most of the recommendations of the consultant who was thought so vindictive. The employment of a medical director was considered wise.

After delivering its message, the committee disbanded and declined to make any public statement on the ground that it would hinder its future usefulness. Although the chairman of this committee stated that his hospital ("Coalton" Hospital) would not allow some of the "Dockston" medical staff members to practice medicine on his staff, neither he nor his colleagues felt any obligation to correct the situation[45]. It must be recalled that the chairman of the committee had firsthand knowledge about the performance of physicians on the "Dockston" medical staff, for he had been the former pathologist at the hospital. Reportedly, he had resigned because of the improper surgery that had been performed at the hospital.

Considering the quick retreat from the controversy on the part of the medical society, the board, even if it had wished to act, could not depend upon the support of the local medical society. The committee was against applying pressure and diagnosed the situation as a problem of poor communication. Public disclosure of its findings was not allowed. However, whereas the medical societies in the region remained silent, the "Tri-County Association of General Practitioners" issued a statement attacking the administrator for "practicing" medicine. The bond uniting members of the medical profession is a strong force mitigating public intra-professional conflict or meaningful peer accountability and review.

Reluctant Participants

Labor unions, especially the United Mine Workers, had established various group-practice clinics in the region. Although they had struggled for staff privileges elsewhere, they quickly dropped out of the "Dockston" Hospital controversy. When challenged about their public responsibility, the UMW chief of staff replied, "We can't save the whole world." The United Steel Workers proposed to take an activist role in the valley, but their plans were vetoed at the national level. The conflict between local leadership of the United Steel Workers and the United Mine Workers made consolidation of labor strength in the region difficult. Both leadership groups were critical of the medical care in the valley but were unwilling to act either together or singly.

The performance of the "Metroville" University group of experts was similar. They took a look, agreed that there were many underlying problems, but walked away saying that the locals must decide to institute change. Physician dominance seemed more important to the university professors than the actual level of medical care in the hospital. Likewise, the State Commision of Hospitals proposed a review but later declined to even visit the hospital in an apparent decision not to become involved. A

"Metroville" hospital planner helped the board (1964) locate the "big city" consultant and then dropped out of the controversy.

Regional and Subregional Health Planning

The most striking alternative medical care institutions were the two coal miners' clinics (UMW) in the region[46]. The establishment of these clinics had caused many conflicts since their inception in the 1940s. In the first case, the miners marched in the streets and finally threatened to build their own hospital before their physicians were appointed to the local hospital medical staff. In the second case, the UMW threatened not to reimburse any patient care at the nearby "Blacksburg" Hospital, where 50 percent of their patients were either miners or their relatives. The board of this hospital quickly invited the physicians from the UMW clinic to join their medical staff. Both clinics introduced various medical care innovations in order to serve the people's needs better, e.g., nurses were trained to do tonometry examinations so that all older adults could be systematically screened for glaucoma every six months or so while they waited to see a physician.

As in most parts of the country, this four-county region had the full range of voluntary health and welfare agencies and associations[47]. Some of these organizations had joined together to coordinate fund raising through the well-known community chest mechanism. This combined fund-raising mechanism is typically promoted and supported by the "corporate" rationalizers in the local towns. Historically and presently, these voluntary health associations have concentrated upon preventive education, medical research, and supportive services, and they have avoided any direct concern with delivery of "mainstream" medical care.

In the region of the "Dockston" Hospital some citizens and health professionals went a step further. They established the "Coal Valley" Health Services, Incorporated, which combined under one administration various formally autonomous agencies, i.e. homemaker's service, visiting nurse's services, and medical loan closet. Later they added the council for exceptional children, a unit in the "Dockston" Hospital, and finally an outpatient mental health clinic. With one exception, namely the psychiatric unit, this agency did not attempt to create any programs that might influence the delivery of traditional medical and dental care in and outside the "Dockston" Hospital.

Thus, until the middle 1960's there had been essentially two main health-planning movements[48]. First, there was the work of the miners, both in terms of developing two clinics in four-county regions and of gaining medical staff appointments at two of the fourteen hospitals in the region. Second, various communities and voluntary agencies started to combine funding efforts and in a few cases to combine administrations to increase efficiency and coordination. With the passage of numerous federal laws, 1965 was the benchmark for explicit federal entry into local health-planning affairs.

While public officials and voluntary agency people discussed regional health planning under the mandate of the Appalachia Regional Act, the medical care "establishment" (essentially composed of private physicians and hospital administrators) was getting organized to develop programs in response to the regional medical program (1965) which the school of medicine in "Metroville" was coordinating[49]. Although regionalization of medical care delivery was a possible emphasis, this state subregional-RMP chose to concentrate its efforts on heart disease, cancer, and stroke. Therefore, the RMP typically focused on projects in one institution instead of broad

planning to integrate or reorganize medical care delivery.

The program managed to link certain rural hospitals to large urban or suburban hospitals in or near "Metroville" in an attempt to rationalize referral patterns and to encourage more steady use of specialty backup in "Metroville." However, most of the rural-urban links were established between the more sophisticated rural hospital and the practitioners in the urban hospitals of "Metroville." The "Dockston" Hospital and other "problem" hospitals in the region remained outside the sphere of influence of RMP planning.

The Comprehensive Health Planning (CHP) legislation differed from RMP in that there was a requirement stipulating that 51 percent of its board members were to be consumers or nonproviders. After activists in the Appalachia regional program realized that there were no health funds available through this act, they shifted their attention to CHP. As the Appalachia program and the local "war on poverty" groups had attracted mainly Democratic Party leaders, so the CHP program attracted Democratic politicians rather than Republican corporation leaders[50]. Some lower echelon health agency directors and professionals cooperated, but most hospital administrators, hospital board members, and physicians in the region chose to debunk or ignore any efforts to organize a four-county health planning agency.

Throughout the five or six years of planning that followed, few changes occurred as a result of CHP legislation. In addition to the failure to integrate the UMW leadership into the system, the local four-county CHP effort represented a basic cleavage between the local medical care "establishment" and Democratic county and town public officials.

In general, social planning and the

bureaucratic organization of health services were not in harmony with either the social values or the dominant interests of this region. Any social movement to rationalize the delivery of medical services was associated with problems similar to those that occur when Western bureaucratic organizational models and social planning are introduced into non-Western cultures[51]. The introduction of concepts such as impersonality, technical supremacy, and loyalty to some abstraction such as the public interest (comprehensive health planning) was often blocked by traditional attitudes toward authority, social class rights and privileges, and professional autonomy, as well as by a strong dependency upon personal and family friendship.

Although most people in this area shared the notion that laymen and especially politicians should not interfere with medical practice, there was a small group in the population that was willing to question this assumption. Even though federal review bodies criticized the inadequate participation of consumers representing all socioeconomic groups in the CHP group, the lower classes with few exceptions were not yet aware of their new "rights" and were generally unwilling to demand a decisive role in health affairs.

Even if they were aware of the need to participate, lower-class leaders often lacked the knowledge, interests, and skills to participate effectively. More importantly, the weak authority structure or mandate under which planning groups operated quickly discouraged serious participation. What continued to be lacking was the opportunity for lower and middle class participation in the establishment of medical care priorities. With the long history of struggles between the UMW and local hospitals, between the unions, between various hospitals boards and their medical staff, between medical staffs at different hospitals, and between different political jurisdictions, it is not surprising that working-class and middle-class citizens would not choose to expend a great amount of effort to rationalize medical care delivery over this four-county area.

CONCLUSION

The following situations are hypothesized to be related to stagnation in rural health care systems: a) weak self-perpetuating voluntary boards of community hospitals[52], b) community willingness to accept a high degree of upper-class paternalism in health affairs, c) physician dominance in the delivery of medical and hospital services, as opposed to a balance between professional autonomy, administrative rationalization, and public accountability, d) general acceptance of ideologies that emphasize consensus and better communication, as opposed to institutionalized conflict between vested interests[53], and finally e) distrust of lower-middle-class citizens to plan for their own medical care. As Freidson summarizes the situation: "professional dominance creates sufficient problems to require the development of a stronger countervailing administrative management and of a better organized clientele"[54].

Social changes in rural health care are found within the following developments. The social organization of citizens to bargain with professionals is a necessary element to improve existing situations. The labor unions, who have organized for other purposes, provide the most powerful example in terms of financial resources and levels of sophistication. Coalitions between local dissidents (including providers and citizens) and government programs from outside the region have some chance of bringing about social change. The role of federal programs, e.g. the Comprehensive Health Planning Program and Regional Medical Program, are important to review and evaluate in this regard.

But as Bodenheimer points out, regional medical programs suffer from the same idealisms that have marked other private and public health programs[55]. For one thing, the groups and individuals brought to the regional conference table are not markedly different from the existing influentials within a region. External (economic) incentives and "disincentives" are at present too diffusely administered to coerce or encourage community hospitals, health agencies, and health centers to create programs jointly for a region and to modify existing medical care delivery. Because of the lack of a strong articulate and informed consumer voice within decision-making processes, what will occur in the future will resemble what Alford calls "dynamics without change"[56].

The maldistribution of medical care in terms of quality and quantity in backward rural areas must be analyzed in relation to a profit-based economic system[57]. However, just as developing Third World nations cannot easily alter the distribution of goods and services between countries overnight and therefore must attempt to adopt institutional forms which meet their immediate problems[58], so most backward rural areas in the United States resist suburban or upper-class solutions to their medical care problems. To even begin to consider the issue of "who does what for what rewards," citizens must take control of their own institutions.

12. Planning Organizational Response to Environmental Problems

Lest the nurse manager think the political behavior of complex organizations has little to do with the day-to-day management function of nurses in an institutional setting, two concrete examples of nurse managers' involvement in organizational change are offered here. The selections illustrate how nurses planned and implemented positive responses to external environmental problems.

In the first situation, nurse administrators and educators of a six-county area undertook a joint interagency venture to assist new graduates' adaptation to institutional employment. The venture was carefully planned following a needs assessment. The net results were a more effectively functioning group of practitioners, a sense of community among hospitals, and the realization by those involved that an interagency program has distinct advantages over programs undertaken by each organization individually.

In the second selection, nurses in a Veterans Administration Hospital, recognizing new community utilization patterns, revamped their ambulatory care program to meet client needs better. It should be noted how creation of the boundary-spanning advocacy clerk role enabled organizational coordination with other community health resource systems, thereby greatly benefiting the patients served.

Both of these selections exemplify good planning technique; considerable time was spent assessing the situation; people to be affected were involved in the planning function; and evaluation of the new programs was an integral part of the planning process.

Planning is a managerial obligation which nurse administrators sometimes do not take seriously. Fending off day-to-day crises often takes precedence over planning activities, which have postponed payoffs. The effective nurse managers in today's institutions need to plan, not only internal operations, but also boundary-spanning systems of care delivery which make appropriate use of health care resources in the community environment.

AN INTERAGENCY INTERNSHIP: A KEY TO TRANSITIONAL ADAPTATION

By Sandra J. Weiss and Ellen Ramsey

Reprinted with permission from *The Journal of Nursing Administration*, Volume VII, Number 8, October, 1977, pp. 36–42.

The lack of attention to the role transition from student to graduate nurse with its concomitant set of expectations and responsibilities has led to inadequate performance in the professional role, a sense of dissatisfaction and powerlessness in the nurse's work, and a resulting high turnover rate. These observations of a nursing community provided the foundation for a joint interagency venture to stimulate new graduate adjustment during the early months of employment. This project included the development, implementation, and evaluation of an internship for new graduates who were employed for the first time as nurses in California's San Joaquin Valley counties of Fresno, Kings, Madera, Mariposa, Merced, and Tulare.

This internship was sponsored by the Area Health Education Center (AHEC), a federally funded organization created to develop educational and training programs for local health manpower needs[59]. AHEC was a catalyst in bringing interested groups together to combine their efforts toward a joint nurse internship program. The San Joaquin Valley nursing community recognized not only the need for an orientation program specifically designed for new graduate nurses, but the necessity to: 1) reduce duplication of orientation programs among hospitals by combining efforts, 2) produce one high-quality program which could benefit the entire nursing community, and 3) give smaller hospitals the opportunity to utilize specially trained inservice personnel. The Program Steering Committee, composed of representatives of both the Nursing Education Committee and the Nursing Administrators Council, felt that AHEC could facilitate an internship program that would allow for such a system of joint resource sharing.

PARTICIPATION IN THE PROGRAM

Six hospitals participated in the joint internship program. Three were small (under 100 beds), and three were large (up to 452 beds). In order to coordinate activities, a program director functioned as the central integrating figure among resource persons from each of the participating hospitals. The resource persons were selected from nursing inservice on the basis of their expertise and interest in the program. While the director's role was to coordinate the learning activities of the entire program, the functions of the resource person were the following: 1) to serve as a liaison between the intern and the program director; 2) to act as a consultant for the interns in the learning activities of certain of the educational modules; and 3) to assist the individual intern by maintaining open communication with her on the unit.

Sixteen interns took part in the program: four were from small hospitals and the remainder from large instutions. Of the 16 interns, six were baccalaureate degree (B.S.) graduates and ten were associate degree (A.A.) graduates.

NEEDS ASSESSMENT FOR THE NEW GRADUATE

The development of the educational modules in the internship program occurred following a needs assessment of problem areas specific to new graduate nurses. This assessment took six forms:

• Questionnaires to head nurses at area hospitals

• Unstructured interviews with supervisors from participating hospitals

• Review of related literature concerning new graduate role transition

• An individualized problem analysis session with those new graduates who participated in the AHEC internship program

• Input from the Program Steering Committee

• New graduate transition data from the Neophyte Nurse Project of the San Francisco Consortium[60]

Head Nurse Questionnaires

The head nurse questionnaire was given to head nurses from both small rural and large urban hospitals in the area. The written questionnaire collected information on the greatest problem areas experienced by new graduates and the skills the head nurses

perceived as essential for new graduates to demonstrate. The designated areas of need were organizational skills, judgment and priority setting, competency in clinical techniques, and management skills conducive to effective team leadership. Of these needs, the head nurses rated organizational skills and priority setting as the behaviors most essential for adequate functioning on the unit.

Interviews with Hospital Supervisors

Supervisors from participating hospitals were informally interviewed to determine new graduate needs in the areas of medications, knowledge of laboratory values, and equipment. The content from these interviews stressed both the frequently used skills that graduates experienced trouble with and newly utilized skills.

Related Literature

Research conducted on the graduates' problems focused primarily on the different sets of values placed upon the nurse as a student and as a functioning staff nurse. Both Kramer and Smith have demonstrated that role-specific values of head nurses and nurse educators appear to differ[61]. The information from these studies and others[62] provided input to the potential conflicts that the interns might experience during their transition, thereby influencing the development of some educational modules.

Individualized Problem Analysis

In order to meet the specific needs of the new graduates in the internship program, a group-needs assessment took place both at the beginning and halfway through the program. This assessment served to analyze problem areas on which graduates felt they needed to work.

Steering Committee Input

Throughout the development of the proposal and the program itself, the expertise of the steering committee was invaluable. As educators and administrators, the members identified their perspectives of the new graduate focusing on competencies in clinical skills, work organization, interdisciplinary and interservice coordination abilities, and interpersonal effectiveness in the role of nurse.

San Francisco Consortium Data

The content from structured interviews with new graduates working in San Francisco hospitals provided the basis for development of transition packages at the San Francisco Consortium (an association for education and urban affairs, comprised of local universities and colleges). The transition packages developed by the Neophyte Nurse Project of the Consortium confronted three general areas of need: adjustment (personal and interpersonal adaptation to the role of nurse); clinical competency (technical skills and judgment ability); and pragmatics of the hospital (the relationship of the graduate to the hospital system). The packages included videotapes, audiotapes, seminar directions, learning objectives, pretests and posttests. These transition packages were made available through the Consortium and represented nine of the 19 clinical content components utilized in the internship program.

CURRICULUM OF THE INTERNSHIP

As a result of the needs assessment, a 16-week internship was created to provide the graduate nurse with both clinical content and clinical experience. Rotation to diverse experiences occurred throughout the program and was individually planned for each intern to fit the needs of the various hospitals. The rotation included the following:

- Two weeks of hospital orientation
- One week of support services
- Eight weeks of medical-surgical
- Two weeks of a chosen elective service
- Shift rotation, both evening and night
- One week of critical care

All-day workshops for the clinical content took place each Monday for the 16 weeks at different participating hospitals. The educational modules provided at these weekly workshops covered one of three defined areas of need: adjustment, clinical competency, and pragmatics of the hospital. This threefold guideline included both the educational modules from the San Francisco Consortium as well as those specifically designed for the internship program.

PROGRAM EVALUATION

Program evaluation took four forms: 1) an overall evaluation of the effects of a nurse internship; 2) an individual evaluation of the effect of each educational module in the program; 3) a subjective evaluation of the program by participating interns; and 4) a performance evaluation of both interns and controls by themselves and their head nurses. Considering the total program evaluation, the internship does appear to have positively influenced the adaptation of the new graduate to the staff nurse role. Each aspect of the evaluation showed some significant findings which have implications for the new graduate nurse[63].

Overall Evaluation

The overall evaluation of the program asked the following question: Does an internship significantly increase the new graduates' adaptation to the role of nurse? Adaptation was defined by the following:

1. Differences in perceived job satisfaction for control nurses and nurse interns after the program, as measured by Munson's Job Satisfaction Index[64].

2. Differences in feelings of powerlessness for control nurses and nurse interns after the program, as measured by a modification of Seeman's Powerlessness Scale[65].

3. Differences in employee stability for control nurses and nurse interns during and after the program, as measured by an Employee Stability Questionnaire[66].

4. Differences in perceived clinical skill for control nurses and nurse interns after the program, as measured by Benner's Clinical Skills Inventory[67].

5. Differences in role conception for control nurses and nurse interns after the program, as measured by Corwin's Role Conception Scale[68].

As the 16 new graduates were assigned to participate in the program, their respective hospitals also designated the names of an equal number of control nurses, matched with the nurse interns for educational background, sex, and starting date of employment. These control nurses served as a comparison group in the project since they took the same test battery as the nurse interns but received none of the curriculum content in the internship. Both the intern group and the control group were given an initial test battery upon entry to the job, a single test instrument six weeks after their starting date, and a final test battery one month after completion of the internship.

Data from the test batteries showed some interesting findings. Prior to the internship, interns as a group were significantly more dissatisfied than the controls with their on-the-job opportunities to share in the determination of methods, procedures, and setting of goals and to have authority within the hospital (involvement satisfaction). And yet, following the program, the interns increased in this level of job satisfaction to equal that of the controls. In addition, the B.S. interns demonstrated a significant increase in their perception of opportunities for belongingness and warmth in their working relationships (interpersonal satisfaction). This finding was not observed in the control group.

After the internship, interns also displayed changes in their service and professional role conception, indicating that nurses who participated in the program felt a greater alignment to patient care values (service) and a greater identification with the profession of nursing (professional). Again, these changes did not occur for control nurses. While the increase in service role conception occurred for both A.A. and B.S. interns, only the B.S. interns differed significantly from the B.S. controls in their professional role conception.

In contrast to these positive changes within the interns, the control nurses experienced a downward trend in adaptation. The control group demonstrated both a decrease in their sense of power and ability to make change in their working environment (powerlessness) as well as an increased dissatisfaction in their assessment of the fairness of working conditions, job security, and financial rewards (extrinsic job satisfaction). These declines did not occur for the intern group, where in fact job satisfaction was found to increase in relation to both working conditions and the authority and power to share in decisions and goal setting.

While all interns experienced adaptational support from the internship, a major difference in program effectiveness between A.A. and B.S. graduates was seen in the function that the internship served for these nurses. For B.S. nurses, the educational modules appeared to **increase** the **adaptation** of the interns in contrast to the lack of change observed in baccalaureate controls. The B.S. interns increased in their professional role conception, demonstrating a higher degree of identification with the philosophy and values of the nursing profession as a whole. No such finding occurred in the control group. In addition, B.S. interns were significantly more satisfied in their evaluation of working conditions, job security, and financial rewards upon completion of the program. Again this finding did not occur for controls.

In contrast to the B.S. graduates, the educational modules appeared to **maintain a** stable level of **adaptation** for A.A. interns rather than the decline in adaptation observed for A.A. controls. Associate degree nurses who did not participate in the program became significantly dissatisfied with opportunities for job security, financial rewards, and adequate working conditions. Also after the program the A.A. controls perceived themselves as significantly more powerless concerning what happened to them and to others in their working environment. These declines were not seen for the A.A. nurses involved in the internship; rather they retained a stable perception of their power to effect change on the job.

While both B.S. and A.A. interns showed positive findings in adaptation after the program, the modules tended to **increase** the level of adaptation for B.S. interns yet functioned to **maintain** the level of adaptation

for A.A. interns. In direct contrast to their educationally matched control nurses, the program appeared to facilitate the adaptation of interns from both educational backgrounds but in distinctly different ways.

Performance evaluation asked the question: Following an internship, is there a significant difference in the clinical performance of interns and controls as perceived by both the new graduates and their head nurses?

This aspect of the evaluation showed some fascinating findings. Interns demonstrated a higher perception of their performance ability on the unit than did controls, were rated by their head nurses as consistently more competent in their performance than controls, and showed a higher reliability between self and head nurse evaluation than did control nurses.

The significant findings were the following: 1) interns rated themselves as more competent in dealing with the realities of nursing practice than did controls (functioning on different shifts and relating to physicians); 2) head nurses rated interns as more competent than controls in acquiring, performing, and getting feedback on technical skills; and 3) in contrast to controls, head nurses and interns showed a high reliability in their positive evaluation of the interns' knowledge and administration of medications as well as their use of IV, oxygen, and suction equipment.

Each of these findings supports the other aspects of program evaluation which indicated a higher degree of adaptation of the new graduates who participated in the internship than for those who did not.

IMPLICATIONS OF AN INTERNSHIP

Considering the findings of the program evaluation, the outcomes of this joint interagency internship were multiple. First, the development and implementation of the program itself created a greater awareness in both education and service relative to the needs and expectations of new graduate nurses. Nursing educators of both baccalaureate and associate degree programs are planning to provide content jointly for their nursing students which will better prepare them for their role transition from student to staff nurse.

In addition, a stronger community bond between hospitals has formed, whereby the educational resources of larger hospitals are shared with smaller ones that have the need but not the facilities. An internship will continue to be available to new graduates of small hospitals as they join forces with graduates from the larger institutions.

Lastly, and most importantly, the new graduates have gained a great deal from the internship program. The joint participation of interns from different hospitals in the community has done much to dispel the myth of "greener pastures." Participants from private, public, small, and large hospitals have come together to understand that each nurse in each facility experiences similar conflicts, similar stresses, and a similar period of growth. This understanding appears to have provided the nurses with a greater motiva-

tion to "stick it out" and confront their problems in a functional manner.

Even more exciting, the results of this internship have shed further light on the dilemma of the new nurse. Perhaps the concentrated efforts of institutions to increase the adaptation of the graduate should not be the primary issue during their transition. Rather an effort toward prevention of maladaptation seems to be a more fundamental issue appropriate to many of these nurses. For the A.A. graduates in particular, representing two-thirds of the nurses in the project's target group, the internship appeared to significantly prevent the decline in a sense of power and job satisfaction which was seen in control nurses. This finding supports the hypothesis that as a nurse continues in her career, a sense of hopelessness, powerlessness, and dissatisfaction sets in that may increase with time. If an internship functions only to prevent this disillusionment from occurring and to keep a nurse's spirits intact, the cause is a valid and realistic one. Keeping new nurses excited about their profession by helping them to develop strategies for coping with stress and for making change within the limitations of the system is an extremely vital and practical goal. This goal appears to be nurtured and furthered within the framework of a new graduate nurse internship.

We in nursing administration are frequently slow in developing interorganizational working relationships with institutions similar to our own. It is as if the competition for patients in which agencies and hospitals engage is carried over into our relationships with our nursing peers. If nurse administrators do form a community coalition, time is frequently spent complaining about mutual problems without seeking new solutions through a pooling of facilities, educational resources or expertise. Do we think that it is all we can do to manage the everyday prob-

lems within our own institution? Do we leave the boundary spanning to the top administrator? When nurses in leadership positions begin sharing resources with other hospitals, agencies, or educational institutions, they find that solutions to many everyday problems lie outside the walls of their own hospital. Weiss and Ramsey's interagency internship project is just one example of how much can be accomplished through interagency cooperative relationships.

We will see more of such arrangements, as Leavitt et al. explain:

The organization of the future will not do its own thing in a wide-open, passive, and unsullied environment. Like the individual, the organization will inhabit a world densely populated by other, related organizations. The organizational world is becoming crowded, interactive, like the urban world of the individual.

The governmental-industrial-educational complexes of today probably represent only the beginning of such interdependent, interactive arrangements. For in a crowded world, organizations will need to cooperate as well as compete, to set common standards and limits, to form coalitions and combinations of all sorts. The ac-tions of any organization will affect other organizations in much more numerous and complex ways than in the past.[69]

For the neophyte manager, the organizational world promises to be more exciting and more relevant, but less orderly, less reliable, less anchored, than in the past. He will have to tolerate more ambiguity and more uncertainty, both about his own career and about his work. He will have to share power and live with extended negotiation and debate. He will have to initiate social relationships in his search for new and innovative organizational coalitions.[70]

DEVELOPMENT OF A PROGRAM FOR AMBULATORY NURSING CARE

By Gwendolyn J. Buchanan

Reprinted with permission from *Nursing Clinics of North America*, Volume 12, Number 4, December 1977, pp. 543–551. Published by W. B. Saunders Company.

The concept of ambulatory care has received growing acceptance by the health care consumer and is identified as the trend of future health care delivery systems. Theories of continuity of care, primary care, community and consumer participation in health care policy, and the importance of preventive care are becoming a reality in the vertical care of patients. Ambulatory care has become an integral part of the health care continuum, with growing emphasis on locally accessible preventive and therapeutic models that offer a comprehensive range of services for all family members.

It is essential that the health care needs of the community be identified and that the consumer be given the opportunity to become involved in the policy making and development of ambulatory health care services. There must be a means to ensure that services are of high quality, comprehensive, continuous, financially feasible, and easily accessible. And, by no means least important, the service must reflect concern for the dignity of the patient.

This paper deals with the history of ambulatory care and the development and growth of a hospital-based ambulatory care program in a Veterans Administration Hospital in a southwestern health science center.

A CONCEPT REBORN

Historically, the concept of ambulatory care is neither new nor original. Soon after the founding fathers of our country declared their independence, health care for the vertical patient was established. In 1786 America's first dispensary was opened in Philadelphia, 35 years after the opening of the first hospital in the same city[71]. Not until the eighteenth century in Europe and the nineteenth century in America, however, did hospitals consider it reasonable to have departments for outpatients. The hospital clinic patients like the inpatients, were expected to be poor. These clinics were generally located in hospitals that were involved in or associated with the education of physicians.

In the early 1900s school clinics were organized in the United States and industrial clinics were established following the first legislation for workmen's compensation. In 1880 the Mayo brothers originated the first group medical practice. The voluntary Visiting Nurse Association was founded during this era and continues today as a vital community health care resource. Under the Public Works Administration of the 1930s and the Hill-Burton Act of the 1940s, health centers were built for local health departments in rural counties and public health agencies in large cities. Not until the 1960s was the comprehensive health center idea reborn.

Recently, nurse practitioners and nurse clinicians have begun to establish group practices to provide health teaching and health maintenance in organized ambulatory care services. This professional nurse service has been well received by the health care consumer and appears to have the potential of expanding progressively.

The Health Maintenance Organization Act of 1973 (Public Law 93-222) proposed a plan for establishing a system of private, comprehensive, prepared health care programs as an alternative to conventional health care systems. A typical HMO assembles a number of health services together for its enrolled members and collects a lump sum or annual fee from subscribers instead of charging a fee for each service. An example of this plan is the Kaiser Foundation Health Plan. Ambulatory care is a principal service of the HMO. This conceptual model has been widely accepted by the consumer and appears to be a trend for future health care delivery systems.

THE VETERANS ADMINISTRATION AND AMBULATORY CARE

The Veterans Administration has maintained a progressive role in this nation's health care delivery system. Recognizing the new patterns of consumer utilization of health care services, the Veterans Administration began to explore some of the problems and possibilities relating to the rapidly changing picture of its outpatient activities. Dr. Paul Haber, in his "White Paper on Outpatient Clinic Operations," March 1, 1971, stated, "The outpatient services in the

VA are a vigorous and growing modality of care, which the VAH has reason to be proud of, and which obviously are going to expand further in the near future. As the cost of hospital care rises, alternatives must be sought, and the prime alternative is, of course, outpatient care.

Nursing and Ambulatory Care

Nurses, like other health care professionals of the past, viewed the outpatient service as a distant, vague care source that existed but was unrelated to inpatient service. This sharp division was apparent even when inpatient and outpatient services were housed in the same institution. Rather than a care continuum, there was care conflict. Nurses in the outpatient service functioned chiefly as traffic controllers, sorters of reports, and technical shot givers. Long lists were maintained and checked, patients were directed from one appointment to another, and a technical, standardized type of performance dominated nursing service.

In 1969, the outpatient nursing service of the VA Hospital in Oklahoma City functioned very much like any other clinic service in other parts of the country. The nursing staff were principally involved in getting the patient through the system. To bring about change, the clinic nursing staff had to examine their philosophy of care in relation to what nursing services were available to the patient during his clinic visit and how well these services were being rendered.

Through the years, the routine of getting the patient through the system during his outpatient visit had become so rigid that it might be compared to directed movement through a maze. Each of the patient's stops, beginning at the reporting desk, had a designated activity that had to be completed before movement to the next. Any deviation from the established route created crises in the system, with the frequently heard comment, "But we've always done it that way." When, at last, the patient arrived at his assigned clinic, the nursing personnel weighed him and measured his vital signs. This information was religiously recorded in the patient's clinical record. There was no evidence of other assessments, if they were being made, since nursing progress notes were infrequently recorded. Patients had complained that they were "routed" like things, not people.

The clinic nurses agreed that if receiving the patients was as ritualistic and lacking in personal exchange as was documented in the clinical records, the patients had well-founded reasons for their complaints. Responding to the increasing dissatisfaction of the patients, the nursing staff worked to make an objective study of the patient's visit, his reception and care. As a result of this evaluation of the flow and movement of patients through the system, the nurses were able to accept the fact that changes in the environment, as well as in the nursing staff, must take place to bring about the shift from things to people.

THE BEGINNINGS OF CHANGE

Problem Solving and Planning

The first task that confronted nursing was to define clinical nursing practice as it related to this setting in this system. Meetings were scheduled to provide the outpatient registered nurses the opportunity to solve problems together. Each nurse was assigned a subject for a literature search to identify current concepts as they related to clinical nursing practice and ambulatory care nursing. She then presented a review to the nursing staff.

The process of reaching group agreement

was painful at times but proved to be constructive. The areas of responsibility in nursing practice were identified and agreed upon by the professional nurses in the clinic. These nursing responsibilities are:

1. Patient counseling
2. Health maintenance
3. Preventive care
4. Primary care
5. Patient education

The next problem was to evaluate present practice. Responsibilities for stocking, stamping, and completing laboratory and radiology requests, scouring sinks, and cleaning doctors' offices were among the many non-nursing activities carried out by nursing personnel. It was difficult to accept that few of the activities performed by nurses in the clinics could be related to clinical nursing practice. However, recognition of the deficiencies in practice served as the impetus for change. The staff were united in a determined effort to develop quality nursing practice in an ambulatory care nursing service and to change the image and role of the outpatient nurse.

Staff Collaboration for Outpatient Care

The clinic nursing staff recruited the help of the inpatient nursing staff to assist them in identifying their role in continuity of care. What, in fact, were the inpatient nurses' expectations of outpatient nursing? Problem solving together brought many new realizations. The combined nursing staff found that they shared many problems as well as common objectives for the improvement of nursing practice. They recognized that their major differences were not in patients and patient care, but in the **environments** in which they provided care. The line dividing inpatient and outpatient services began to crumble. The patient, for the first time, became the central, most important factor in planning a continuum of care.

Several benefits of the combined nursing staff meeting were identified. This practice continues today with combined staff nurse inservice programs, head nurse meetings, and clinical supervisor meetings. Planning and problem solving together has enabled a smooth transition of the patient from inpatient to outpatient care and from ambulatory care to inpatient care. There is mutual pride in the effectiveness of the overall health care program.

Strengthening Outpatient Care

While the program changes were planned, the outpatient census continued to rise as congressional authority was granted for increasing segments of outpatient care in VA hospitals. With growing numbers of veterans making application for treatment and the admission area being the only entry to the health care system, this area was given first priority in the organizational plan. Faced with the dilemma of maintaining quality care, identifying the true emergency, and providing services for the awaiting consumer, nurse practitioners were selected to triage all applicants and establish priorities of care.

Concurrently, along with efforts to collaborate with inpatient nursing staff, the outpatient clinic and admission nursing staff worked with the physicians, dietitians, and social workers to strengthen the multidisciplinary team concept. Team conferences were held on a regularly scheduled basis to permit group problem solving in planning for changes in the care system.

Through the first phases of establishing this concept, the team members met together for one hour each week. With group prob-

lem solving, each discipline acquired a deeper understanding of the services and responsibilities of the other team members. They learned to work together and plan together and gained insight in how best to utilize the talents and strengths of each discipline in order to improve health care services.

As the team developed and acquired operational strength, the frequency of the formal meetings was reduced. Today the multidisciplinary team concept is the only approach accepted by the dedicated staff. The formal meetings have been reduced to a monthly and/or on-call basis to plan and solve problems together as new challenges present themselves.

The clinic staff was determined that the holistic approach would dominate in the delivery of nursing care and that the image of the role of nurses could change. Staff members were appointed to various groups that were charged with the responsibility to develop programs to refine assessment skills, revise outpatient nursing procedures, restructure nursing personnel assignments, and structure patient education programs. Giving up rituals of the past was difficult. Eliminating unnecessary lists and logs met with resistance from physicians and administrative and nursing personnel. With their eagerness to put the proposals into practice, the nurses had not identified all of the emotional attachments to the many lists and books. Symbolically, these "things" represented control, stability, and reliability. Clinging to these documents continued even after change.

CURRENT HEALTH CARE DELIVERY SYSTEM

Change to a Modular System

Through this decade of growth and change, all Oklahoma City VAH health care providers and administrators were evaluating their roles in the delivery system of vertical care and upgrading services to better meet consumer demands. The chief of ambulatory care, a physician, made many progressive changes in the delivery system procedures during the same period that nursing was revising nursing practice. One of his most innovative changes was organizing the clinics into five modules.

Each of the existing 49 specialty and subspecialty clinics was assigned to an administrative unit, labeled A, B, C, D, or E. Each unit or module, which was made up of 5 to 10 clinics, assumed the total responsibility for scheduling of appointments of the clinics assigned to it. A stable multidisciplinary staff was appointed for each module. Each patient was assigned a "home" module where he would always report. This supported the belief of the staff that if the patient had a team he could relate to and faces he was familiar with, he would feel a more personal relationship with providers of service. The patient's phone inquiries concerning his clinic visits were directed to his "home" module where he knew the person answering his questions.

There were many benefits from this change, but one of the most observable was the reduction of long lines at one central receiving unit for both the entry and exit of the patient. This model system has proved to be cost effective as well as supportive of efforts to provide more personalized service.

Nursing under the New System

A registered nurse team leader is assigned to each module. In this important leadership role, the nurse directs the nursing personnel on the team and collaborates with the physicians and other service personnel working in the module. The nursing team leader is expected to develop structured patient education programs, provide unit inservice,

assume responsibility for maintaining quality nursing care, and act as a liaison to the head nurse of the clinics.

The registered nurses have assumed responsible positions in both primary and preventive care in this multidisciplinary approach to health care. The ritualistic reception of the patient has been changed, and nursing completes an initial assessment and/or evaluation of progress for each patient reporting to the clinic. Flow sheets, assessment guides, and other tools have been developed as idiosyncrasies of each clinic population are identified.

Patients receive individual instructions for any special procedures such as barium enemas, gall bladder series, and so on. Nursing has become an important member of the team in minor surgery and provides pre- and postoperative patient education and counseling. The professional nurses in each of the clinics are available to give the patients instructions concerning their medications and their treatment regimen. The nursing team, consisting of registered nurses, licensed practical nurses, nursing assistants, and health technicians, work cooperatively to ensure continuing comprehensive nursing care for all patients in the ambulatory care clinics.

Efforts to structure patient education programs were productive. Multidisciplinary teams completed patient education programs for three clinical specialties, and regularly scheduled classes are being held for both inpatients and outpatients and their families. Patient education is recognized and accepted as a treatment modality. Patients are scheduled for classes in the same manner as with scheduled clinic visits. Class participants have expressed their appreciation for the opportunity to learn to care for themselves.

Growth of Programs and Problems

A continuous pattern of rising patient census in ambulatory care has occurred, as shown in Exhibit 3-2. As the daily census continues to grow, so do problems related to space, staff, and organization of service.

Compounding the identified problems, the consumer may have experienced getting through the system are the number of miles he may have traveled and hours he may have spent in reaching our hospital-based ambulatory care setting. The patient may travel as far as 362 miles from his home town to report for his clinic visit. Travel is frequently by bus or car, but may be by ambulance or Medivan. The nurses had to develop an awareness of the effects on patient care planning when the patient has to travel a long distance. For example, the time that the nurse takes to instruct the patient to take a prep laxative or to administer his prep enema needs to be taken into account. For the traveler, it is important that the clinic visit be as efficient and productive as possible and that the consumer, the veteran, receive quality services for his identified health care problems.

Coordination with the Community

Equally as important is the vehicle for communication from the VAH to the veteran's home town health care resources to ensure a continuum of care. The professional nurses initiate nursing care referrals to public health nurses, Visiting Nurse Association, Indian Service, Migrant Service, and other available nursing services in the veteran's community. Services requested, based on the individual patient's needs, may be for continuous nursing care, patient and family health education, counseling, or home evaluation to assist in planning patient care. The nurses collaborate with the physician and members of other health care disciplines to plan for the patient's uninterrupted care.

To ensure an open communication line that allows the veteran to call for assistance with his health care problems, the position of coordinator of information was created. The advocacy clerk, a nonprofessional, is available by phone 24 hours a day to help

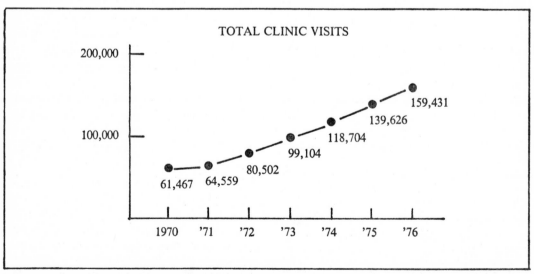

Exhibit 3-2.   Clinical Patient Populations from 1970 through 1976. (Veterans Administration Hospital, Oklahoma City, Oklahoma.)

veterans with many problems. She acts as a phone liaison for the veteran to the many services in ambulatory care and inpatient services. The veteran may have questions about his treatment, medications, equipment, and other care procedures while he is at home. He can call the advocacy clerk, who will contact the responsible service and respond to the patient's questions. Each day the advocacy clerk also calls veterans who live alone, to see if they are doing well and if they need help. This unusual and valued service benefits the veteran and is another resource to ensure continuous care.

SUMMARY AND CONCLUSIONS

Health care delivery systems that provide care for the vertical patient are increasing in numbers as well as in the scope of services that they provide throughout the nation. The nurse's professional relationship with other health care team members has broadened, increasing areas of overlapping responsibilities and roles in these health care settings. A more highly educated population is increasingly demanding as to the prerogatives and special position of the profes-

sionals who serve them. Change is the password today, and nurses must move together to meet change, ensure quality, and maintain an awareness of consumers' health needs. Nurses, like other health professionals, are faced with the pressing need to critically reevaluate their roles in the delivery of ambulatory care.

The effectiveness of the strategies to bring about change in the Oklahoma City VA ambulatory care nursing services will become apparent as the consumer response to new programs is evaluated. While it is too early to forecast the ultimate outcomes, the nursing staff reflect increased satisfaction in their clinical practice, and the patients and their families have been enthusiastic in their involvement in the patient education programs. A close collaborative relationship with the state, county, and local health resources is well established. A strong multidisciplinary team works together effectively in planning and providing care. Non-nursing duties once performed by the outpatient nursing staff have been relinquished to the appropriate service. The registered nurse has become a "professional," accepting the responsibilities of the role in clinical practice and the accountability of performance.

James Thompson's propositions are well illustrated by Gwendolyn Buchanan's description of the development of a new VA ambulatory care program. The rigid centralized patient admission system being used was inadequate. A new program was needed which would allow for both the expansion of ambulatory services and for increased numbers of patients. In response to the need for increased differentiation of the hospital's boundary-spanning units, the ambulatory clinic was divided into five modules; each module responding to its own segment of the task environment (veterans at home being followed by the clinic). In addition, new

coordinating people were used to direct the veteran to his specialized boundary-spanning unit.

Ritualistic procedures were replaced by the attentions of human beings who could respond sensitively to the series of decisions a patient problem frequently requires. The VA used specialized boundary-spanning units to accommodate the hospital to its task environment. The net result was a flexible system which allows for increased expansion efficiency, assured patient follow-up, and offers more effective patient care. The change agents used effective planning technique.

We have discussed how organizations accommodate to environmental problems by developing boundary-spanning components, by engaging in political action, by utilizing interagency networks, and by planning institutional change. The aim of this volume on organization-environment relationships has been to press nurse managers to look out at the community, recognize environmental barriers to organizational functioning, and plan actions which facilitate organizational accommodation to a probabilistic environment.

Central to the method of accommodation which the nurse manager may choose is the activity of health planning. A detailed description of the technology of health planning is beyond the scope of this anthology. It is important, however, for institutional administrators to understand the unique characteristics of health planning and be apprised of how they can plan their institution's resource development within the larger context of the areawide health care system: the external environment.

John D. Thompson[72] defines **planning** *as the forecasting of the optimal achievement of selected objectives per unit of available resources.*

Planning methodologies generally follow six steps, regardless of the size or scope of the planning endeavor. They are:

1. *Definition of an overall good*
2. *Breaking down the goal into objectives (subgoals)*
3. *Discovering alternative courses of action for achieving goals*
4. *Choosing between alternative courses of action to arrive at a plan*
5. *Implementing the plan*
6. *Evaluating the results (outcomes)*

Health planning follows this same general planning format, but differs from many other types of planning by virtue of the high degree of indeterminism which characterizes health service data and collection methodologies. As W. Shonick explains:

Many different sorts of activites are called planning, all with some justification for there is no authoritative definition of what types of administrative activities constitute planning and what definitely do not. This open-ended quality is the underlying feature of nearly all aspects of social planning and health planning in particular. Decisions are constantly being made despite uncertainty in response to problems that are either explicitly formulated or implicitly sensed.[73]

One reason for the high degree of uncertainty in health planning is that the state of the art in health planning is rarely able to link health status (outcomes) directly to health resource development. This leaves the planning process open to political persuasion, and in the face of today's scarce health resources, draws health administrators into boundary-spanning activities and transactions with groups and agencies in the institution's external environment. Shonick defines health planning as "almost any kind of activity that proposes methods for operating systems or subsystems of health care delivery"[74].

*Health planning projects differ in sponsorship and purpose. John Thompson describes three models. First, **institutional planning** has as its purpose the optimization of institutional objectives. It is usually sponsored by the institutional board or people within the institution, and the planning is most often concerned with forecasting the market. Institutional planning tends to be a competitive closed-systems model. The second model of health planning, **program planning,** is used to maximize a particular problem-solving process through application of cost-benefit analysis. H. V. Berg's description in this volume of how a nurses' aide training program was developed is a good example of this type of health planning. The environment on which program planning is based is considered a passive factor that can be manipulated in a predictable way. Thompson states:*

Although program planning is more often than not multi-institutional in scope, and therefore may pay more attention to institutional coordination than corporate (institutional) planning, it still tends to limit itself to the solution of one, two or three problems.[75]

*   **Community planning,** the third model of health planning, differs from the first two types by starting with a population base within a defined geographical area or a region. The primary purpose of this type of planning is "to set priorities, make recommendations for actions within these priorities, and promote implementation of these recommendations with the goal of improving the overall well-being of that population"[76].*

*   Community health planning is also called areawide health planning, regionalization, or comprehensive health planning, depending upon the context and scope of the activity. Program planning and community planning are open-systems models which require some degree of interagency negotiation and cooperation. Community planning is definitely a political activity that ultimately reflects environmental resource limitations and the social values of the groups of people living in the community. For these reasons, the planning decisions in community health planning are usually based on detailed analysis of the health needs of the community,*

*and a wide variety of people participate in the decision processes. Robert Mac Stravic's book **Determining Health Needs**[77] is an excellent resource book for nurse administrators who take an active part in community health planning. Mac Stravic describes how to accurately identify client populations, predict health needs, and evaluate predicted futures on the basis of preset standards or by examination. The open-systems model of community health planning represents a continuing trend away from the market as the proper vehicle for ensuring optimal health services and towards a rational or political planning model.*

The next selection is a description of community health planning in an Ohio community. The planning was limited to the mental health component of an areawide health systems plan. Nevertheless, a broad assortment of professional planners, public officials, private health practitioners, health institution administrators, and community lay people participated in the planning processes. The layering of the committee structures made it possible for these diverse groups to interact and exchange information, thus allowing integration and plan development. Note how one of the most difficult parts of the process was getting health professionals to relinquish cherished assumptions which could not be substantiated. The planners were careful to use an open process to ensure cooperation of the medical professionals and institutional administrators who control the data essential to the process.

Nurse administrators are increasingly being recruited for community health planning boards and committees. It is our opportunity to assure quality health care for populations as well as for individuals and patients who use the health care institution which employs us. The boundary-spanning role of health planning is a mechanism which makes it possible for us to plan our own institution's resource development in cooperation with other agencies and organizational networks in our institution's external environment.

PLANNING FOR MENTAL HEALTH SERVICES

By Thomas M. Wernert

Reprinted with permission from *Administration in Mental Health,* Volume 6, Number 3, Spring 1979. Copyright © 1979 by Human Sciences Press, 72 Fifth Avenue, New York, New York.

Community mental health services in Ohio are under the aegis of county mental health and mental retardation boards. These boards are basically responsible for the review, evaluation, coordination, planning, and funding of services provided by local agencies.

In 1977, the Lucas County Mental Health and Mental Retardation Board and a previously designated Health Systems Agency agreed to develop mutually the mental health component of the health system plan (HSP) to be submitted to the Department of Health, Education, and Welfare (HEW). A health system plan is required by HEW before final designation as a health systems agency[78]. Both agencies agreed on a planning process that included the establishment of a joint committee of the Health Systems Agency and the County Mental Health Board. This committee was composed of lay people representing both agencies as well as provider and consumer interests, and it was staffed by a professional from each agency.

A MODEL

To provide a theoretical and practical basis for beginning the planning process and the actual data collection, a planning model was developed. It contained the five following components:

1. **Demand.** The emphasis was on analyzing utilization data to determine client characteristics and distribution of existing cases by census tract. The generation of an aggregate picture of the sociodemographics of the total system was the goal.

2. **Occurrence.** A survey instrument was administered to a random sample of the adult population of the county to determine the latent need for service. The instrument generated specific data that were used to project possible risk groups.

3. **Risk forecast.** Risk indicators were isolated from the needs assessment instrument and then used to develop a model for forecasting future consumer needs for services. The model was used to project actual numbers of people who, by 1982, would require some type of mental health service.

4. **External forecast.** Social, political, and economic issues as well as movements and changes were explored as to their potential impact on need, demand, and population trends.

195

5.   **Priority.** The last task was the analysis of the information generated from the preceding components and the development and prioritization of goals and objectives.

The model provided the parameters for data collection and guided the planning staff in a consistent collection/analysis process. For this discussion, the components are compartmentalized and follow in an orderly fashion. In actual practice, the generation of data, return of surveys, analysis, etc., was an ongoing process within each component. The utilization datum for the private hospitals, for example, was one of the last items to become available to the committee. Although its review fell within the demand component, work on the other components had proceeded. New data were continually becoming available, but the model was flexible enough to allow new input and subsequent modifications.

DATA COLLECTION

The data collection was handled by specifically assigned staff, and there was an attempt to keep this process separate from the actual data analysis and plan development. There were three simultaneous collection processes: (1) data on client/patient-related items such as residence, age, and diagnosis of individuals served by the mental health network; (2) statements on needs, priorities, and gaps in service from all private agencies and practitioners and from key community people in policy-setting and/or influencing capacities; and (3) communitywide perceptions of needs and services generated from a countywide survey. The process, problems, and challenges of each of these three collection processes will now be discussed in detail.

Utilization Material

The Ohio Department of Mental Health and Mental Retardation uses a client-based admission and termination system that was easily accessible to the county board for the planning p ocess. Detailed data for county residents were collected from the state system. A yearly analysis by clients and by service providers explored the following information: census tract, age, sex, diagnosis, race, and disposition. Disposition is a series of descriptive codes that indicate the level of success around treatment and the referral pattern for continuity of care. The analysis, for example, looked at a specific code that related to clergy in an effort to study links between formal systems and the group of informal care givers.

A similar analysis, utilizing the departmental data network, was carried out for the local state hospital. The hospital provides services to approximately 5000 people a year, and had an average daily census of 550 to 600 patients during the years studied (1974–1976). The data analysis was conducted on all individuals formally admitted in addition to the individuals served at the hospital's admitting, consultation, emergency (ACE) unit. This particular unit of the hospital can hold an individual for emergency intervention for 72 hours and provides services to approximately 1500 people per year. Although the information from this source was readily available, it was not broken down by census tract, which was necessary for area analysis and forecasting.

The solution to this problem was the temporary assignment of a clerk to work with the hospital medical records staff to collect manually the addresses of all admissions and terminations. The addresses were then converted to the appropriate census tract through an address-matching program that was available at the University of Toledo's computer center.

This was the only time that specific identifying information was collected. The addresses in this case were not kept and were destroyed immediately after they were used. The clerk collecting the data was not allowed access to any file, but obtained the data from the hospital log, which contained only specific demographic information.

Another major source of utilization data was private hospitals and other agencies not connected with either the county board or the Ohio Department of Mental Health and Mental Retardation. Within the planning area are three hospitals that provide psychiatric services and approximately ten private agencies that offer mental health or mental health-related services. The Health Systems Agency played a vital role in requesting and obtaining the participation of all three hospitals. The data returned, however, varied in quality and quantity; two of the hospitals provided generalized data by zip code, age, and sex; another hospital, which had a sophisticated computer system, supplied census tract, age, sex, race, diagnosis, and length of stay. The hospital with the computer network is the largest private provider of inpatient services for the area; this increased the reliability of the demand information used in forecasting.

The private agencies presented a greater challenge because their data systems were manual, and available aggregate data were not easily compared with other existing data. Through the efforts of the United Way, a large family service agency is participating in a study to develop a mechanism to address this issue and is now collecting census-related information by client. These data will not be available for use in the initial analysis, but will be used in the yearly update process.

The volume of the data used varied, particularly in the private, nonprofit agency sector such as the private hospitals and agencies within the United Way system. Generally, however, the overall material generated supplied an accurate picture of demand and utilization for all major mental health providers.

Survey Material

In addition to the data just described, private providers, private practitioners, and key community people were surveyed. The survey instruments and methods were excerpted from a recently published National Institute of Mental Health monograph, **Planning for Change: Needs Assessment Approaches**[79]. The document describes the key informant and community forum activities as follows:

1. **Key informant.** This method is a research activity based on information secured from those in the area who are in a good position to know the community's needs and utilization patterns.

2. **Community forum.** It is similar to the key informant approach in that it is based on the views of individuals; some of the more serious disadvantages of the key informant approach are reduced, however, by widening the circle of the respondents. Forum studies are designed around a series of public meetings.

The survey instruments were utilized in a slightly different manner than suggested within the document. The key informant interview was mailed to the director of every mental health agency or program and to each private practitioner serving the study area. The community forum instrument was personally administered to major policy makers and was used to summarize several large meetings with groups of interested professionals. For the purposes of the forum instrument, the major policy makers included the county commissioners, local mayors, heads of public agencies such as welfare and Social Security, and all major insurance agency heads.

The survey administered to practitioners was mailed to each individual listed in the local phone directory under the following categories: marriage counselors, psychologists, psychiatrists, and social workers. There were approximately 85 people so identified; this initial list was expanded to approximately 100 individuals prior to final mailing. The return rate, contrary to what national trends are presently indicating, was extremely high; approximately 80 percent of the individuals surveyed returned usable surveys. A breakdown of categories indicates that psychologists, social workers, marriage counselors, and counselors had a 90 percent return rate. This rate dipped to approximately 60 percent for private psychiatrists. The overall high rate of return is attributed to an aggressive follow-up procedure that included a written reminder within three weeks followed by a series of three personal telephone requests.

Generally, the surveys attempted to tap perceptions of need, levels of existing services in the case of providers, and a request for the respondent's personal priorities. The information returned was general in nature because of the general nature of the questions. The following are examples of survey questions:

1. In your opinion, what are the most pressing mental health needs of those living in the community?

2. How would you characterize the clientele of your organization?

3. What should the priorities for mental health services in this community be?

Community Needs

After a review of the needs assessment literature[80] and a discussion with staff of the Biometrics Division of the National Institute of Mental Health, the Florida Health and Family Life Survey Instrument was chosen as the instrument for a community-wide needs assessment. This choice was made because those involved in the Lucas County study believed, as do Warheit, Bell, and Schwab[81], that "surveys are the most scientifically valid and reliable and, hence, the most useful of all needs assessment approaches. . . . " The instrument chosen contained approximately 150 questions; however, ten new questions pertinent to county needs, including items on housing and use of the informal care systems, were included and a few items were excluded. Because the questionnaire uses specific questions to generate risk scales, care was taken not to remove any of those items.

Specifications for conducting the survey and analyzing the results were then developed. The specifications for conducting the survey were as follows:

1. A sample size of 500 with a valid replacement procedure in the event interviews were refused or not completed. This sample size assured a 2.5 percent level of accuracy.

2. A random probability sample controlled by age, sex, and race.

3. Households were to be randomly selected within blocks or census tracts.

A local marketing firm submitted the lowest acceptable bid and a contract was negotiated that further strengthened the procedure by limiting the number of surveyors to a small constant group, ensured ongoing supervision, and mandated that all interviews would be conducted on a face-to-face basis. Interviewer escorts were provided in up to 20 percent of the geographic areas considered to be high-risk areas. The marketing firm suggested presenting each individual interviewed with a silver dollar at the end of the session, which averaged one and one-half hours. The survey of the entire county took approximately three weeks and within one month after the interviewing, a 100 percent verified clean tape was available for analysis.

The result of this process was a sample of 493 respondents selected randomly and proportionately to the four catchment (service) areas in the county. The interview instrument contained 144 questions and was administered to persons 17 years or older. Included in the instrument were biographical, demographic, social, and occupational data and a medical history, perceptual indicators of physical and emotional health status, and behavioral information for each respondent.

A QUESTION

As the analysis of the data and the planning proceeded, an interesting dilemma arose. The dilemma was to decide if existing planning and operating assumptions or the generated data should be used to provide the basis for planning. The participants chose the data as their only planning base. This in turn required that they question and eliminate many ingrained assumptions. The professional members, for example, could not assume that the exclusive use of one professional group was necessarily vital to a healthy community. The process of developing goals and priorities for a large, complex mental health network and then withstand public scrutiny required that the final product be based on the data[82].

Although the mental health field has placed a great deal of reliance on subjective assessment, the lay committee supervising this process would not operate under those assumptions. Since the plan was a five-year forecast with yearly updates, the decision was made to delay inclusion of any unsubstantiated item, require its analysis between the updates, and then make recommendations based on the analysis. This was not an easy process for the professional members of the committee, including those staffing it. An example of this is seen in one of the member's recommendations to increase training resources for a particular category of professional. This member felt strongly that the reason that more mental health services were not delivered and were not at the quality level that this individual expected was that this particular profession was not readily available in the community. The data did not demonstrate that an increase in this particular profession would better meet the identified needs of the community; the recommendation was, therefore, not defensible with the data on hand and, therefore, was not included in the plan.

It is important to reemphasize the need to base the plan on the data. In some cases it confirmed long-standing assumptions; in many cases it did not. The data clearly demonstrated, for example, the wide use of general medical practitioners for general mental health problems, but it dispelled the belief that a lack of transportation in the county presented a barrier to service.

PLANNING DATA GAPS

It is also important to note that there were major gaps in the available planning data. The first, and certainly the most serious problem is the issue of the private practitioner. An analysis of the reimbursement levels of a major insurance carrier indicated that many people were receiving mental health services from private practitioners, primarily private psychiatrists. Initial attempts to obtain the information through insurance companies did not work; they were unable to provide client-based socio-demographic data or even levels of reimbursement. Inroads to the private practitioner and the professional associations must be made if significant information is to become available. To address this deficiency, the planning committee has developed an objective to establish a task force that includes the local Academy of Medicine, the local branch of the AMA, other professional groups, and teaching institutions to suggest solutions for generating these data.

The second major data gap was with the private agencies. In most cases, these agencies are connected with the United Way, which serves the three county areas surrounding metropolitan Toledo. The United Way, after the initial formation of the committee, was included in the planning process with the addition of a board and staff member to reflect adequately their important contributions to mental health services and facilitate the cooperation of their agencies.

This gap is primarily reflected in an ability to capture at a client level the activity of the four family service agencies and five or six specialized agencies. Initial indications from aggregate data demonstrate that these agencies are the major providers of mental health care to middle-class families and are a vital link in the mental health care network. Several elements may exert a positive influence on the collection of more detailed information in the plan's first update: (1) the active involvement and enthusiasm of the United Way representatives, and (2) a new state law that requires that any agency seeking mental health reimbursement under existing group insurance plans must obtain a certificate of need, either from the County

Mental Health Board or the Health Planning Agency.

In both the case of the private practitioner and private agencies, the problem of data retrieval seems to be remediable. The United Way continues to work with its agencies, and there is hope that the local professional groups will be able to influence their members.

The third major gap in data, however, presents a technological question that may be more difficult to resolve. The utilization of emergency services for mental health problems in both the general health and formal mental health network is on the rise. These cases and the crisis calls now being handled on the 24-hour emergency phones present a unique data retrieval issue; the process must develop some mechanism to analyze the clients and the expressed needs being handled at this level.

RESULTS OF THE PROCESS

The plan is now being submitted in draft form for public review and comment; the individuals involved have discussed the political ramifications of the items that most dramatically affect the consumer/provider communities. There are recommendations in the plan that many of the committee members regard as being new and innovative and that question the assumptions under which many mental health care providers have been operating. All of these items are likely to generate a great deal of discussion and, potentially, some hositility.

The committee members continually return to their original tenet—the need to back up all assumptions, goals, objectives, and strategies with data. When the plan was submitted for public review, the committee clearly indicated that the draft could and should be questioned by any individual or group. They will have the opportunity to modify the plan; any such request, however, must be substantiated by defensible data. The committee feels comfortable; it perceives that the plan is reliable and valid, considering the information that is available.

CONCLUSION

This article detailed some data collection methods and procedures and suggested a model for using these data. Some of the more pertinent suggestions are summarized as follows:

1. A well-organized, open process seems to ensure the cooperation of the individuals and groups who control the data that any planning process requires. The planning effort should be well publicized and should include a broad, representative segment of the providers. Specifically, a mechanism such as a newsletter or a series of mailed planning updates should be established to allow a continual flow of information. Correspondingly, this mechanism should include a method that allows feedback on these updates to ensure two-way communication.

2. Although the planning process is vital, lay involvement, "buy-in," and leadership are equally important. The active, knowledgeable involvement of the committee members provides a healthy balance to the professional and ensures a closer contact with broader community interests and concerns. Any such planning effort should be directly accountable to a group or board composed of lay leadership and representatives of major mental health policy setters.

3. Finally, such an effort must be grounded in data that are scientifically gathered and then analyzed in a method and manner that are theoretically and technically sound. Federal, state, and local resources are presently available to assist in such efforts, and should be sought out and studied prior to starting the project. Without the help of the National Institute of Mental Health, staff from the State Division of Mental Health, and several local university professors, the process described would still be getting underway.

DOING BETTER AND FEELING WORSE:
THE POLITICAL PATHOLOGY OF HEALTH POLICY

By Aaron Wildavsky

Reprinted with permission of DAEDALUS, Journal of the American Academy of Arts and Sciences, Boston, Massachusetts. Winter 1977, *Doing Better and Feeling Worse: Health in the United States.*

Although specific environmental constraints to be considered in planning health care services may vary from community to community, there are environmental constraints common to all organizations which make up the national health system. This final selection describes some macroenvironmental factors, and projects future courses of action for our society. If this volume has succeeded in making managers aware of organization-environment relationships, they will recognize the value of articles such as this, which inform us of social, political, and cultural conditions which affect the viability and future of our health care institutions.

According to the Great Equation, Medical Care equals Health. But the Great Equation is wrong. More available medical care does not equal better health. The best estimates are that the medical system (doctors, drugs, hospitals) affects about 10 percent of the usual indices for measuring health: whether you live at all (infant mortality), how well you live (days lost due to sickness), how long you live (adult mortality). The remaining 90 percent are determined by factors over which doctors have little or no control, from individual life-style (smoking, exercise, worry), to social conditions (income, eating habits, physiological inheritance), to the physical environment (air and water quality). Most of the bad things that happen to people are at present beyond the reach of medicine.

Everyone knows that doctors do help. They can mend broken bones, stop infections with drugs, operate successfully on swollen appendices. Innoculations, internal infections, and external repairs are other good reasons for keeping doctors, drugs, and hospitals around. More of the same, however, is counterproductive. Nobody needs unnecessary operations; and excessive use of drugs can create dependence or allergic reactions or merely enrich the nation's urine.

More money alone, then, cannot cure old complaints. In the absence of medical knowledge gained through new research, or of administrative knowledge to convert common practice into best practice, current medicine has gone as far as it can. It will not burn brighter if more money is poured on it.

No one is saying that medicine is good for nothing, only that it is not good for everything. Thus the marginal value of one—or one billion—dollars spent on medical care will be close to zero in improving health. And, for purposes of public policy, it is not the bulk of present medical expenditures, which do have value, but the proposed future spending, which is of dubious value, that should be our main concern.

When people are polled, they are liable, depending on what they are asked, to say that they are getting good care but that there is a crisis in the medical-care system. Three-quarters to four-fifths of the population, depending on the survey, are satisfied with their doctors and the care they give; but one-third to two-thirds think the system that produces these results is in bad shape. Opinions about the family doctor, of course, are formed from personal experience. "The system," on the other hand, is an abstract entity—and here people may well imitate the attitudes of those interested and vocal elites who insist the system is in crisis. People do, however, have specific complaints related to their class position. The rich don't like waiting, the poor don't like high prices, and those in the middle don't like both. Everyone would like easier access to a private physician in time of need. As we shall see, the widespread belief that doctors are good but the system is bad has a plausible explanation. That's the trouble: everyone behaves reasonably; it is only the systemic effects of all this reasonable behavior that are unreasonable.

If most people are healthier today than people like themselves have ever been, and if access to medical care now is more evenly distributed among rich and poor, why is there said to be a crisis in medical care that requires massive change? If the bulk of the population is satisfied with the care it is getting, why is there so much pressure in government for a change? Why, in brief, are we doing better but feeling worse? Let us try to create a theory of the political pathology of health policy.

PARADOXES, PRINCIPLES, AXIOMS, IDENTITIES, AND LAWS

The fallacy of the Great Equation is based on the Paradox of Time: past successes lead to future failures. As life expectancy increases and as formerly disabling diseases are conquered, medicine is faced with an older population whose disabilities are more difficult to defeat. The cost of cure is higher, both because the easier ills have already been dealt with and because the patients to be treated are older. Each increment of knowledge is harder won; each improvement in health is more expensive. Thus time converts one decade's achievements into the next decade's dilemmas. Yesterday's victims of tuberculosis are today's geriatric cases. The Paradox of Time is that success lies in the past and (possibly) the future, but never the present.

The Great Equation is rescued by the Principle of Goal Displacement, which states that any objective that cannot be attained will be replaced by one that can be approximated. Every program needs an opportunity to be successful; if it cannot succeed in terms of its ostensible goals, its sponsors may shift to goals whose achievement they can control. The process subtly becomes the purpose. And that is exactly what has happened as "health" has become equivalent to "equal access to" medicine.

When government goes into public housing, it actually provides apartments; when it goes into health, all it can provide is medicine. But medicine is far from health. So what the government can do then is try to equalize access to medicine, whether or not that access is related to improved health. If the question is, "Does health increase with government expenditure on medicine?" the answer is likely to be "No." Just alter the question—"Has access to medicine been improved by government programs?"—and the answer is most certainly, with a little qualification, "Yes."

By "access," of course, we mean quantity, not quality, of care. Access, moreover, can be measured, and programs toward an equal number of visits to doctors can be reported. But better access is not the same as better health. Something has to be done about the distressing stickiness of health rates, which fail to keep up with access. After all, if medical care does not equal health, access to medical care is irrelevant to health—unless, of course, health is not the real goal but merely a cover for something more fundamental, which might be called "mental health" (reverently), or "shamanism" (irreverently), or "caring" (most accurately).

Any doctor will tell you, say sophisticates, that most patients are not sick, at least physically, and that the best medicine for them is reassurance. Tranquilizers, pain killers, and aspirin would seem to be the functional equivalents, for these are the drugs most often prescribed. Wait a minute, says the medical sociologist (the student not merely of medicine's manifest, but also of its latent functions), pain is just as real when it's mental as when it's physical. If people want to know somebody loves them, if today they prefer doctors of medicine to doctors of theology, they are entitled to get what they want.

Once "caring" has been substituted for (or made equivalent to) "doctoring," access immediately becomes a better measure of attainment. The number of times a person sees a doctor is probably a better measure of the number of times he has been reassured than

of his well-being or a decline in his disease. So what looks like a single goal substitution (access to medicine in place of better health) is actually a double displacement: caring instead of health, and access instead of caring.

This double displacement is fraught with consequences. Determining how much medical care is sufficient is difficult enough; determining how much "caring" is, is virtually impossible. The treatment of physical ills is partially subjective; the treatment of mental ills is almost entirely subjective. If a person is in pain, he alone can judge how much it hurts. How much caring he needs depends upon how much he wants. In the old days he took his tension chiefly to the private sector, and there he got as much attention as he could pay for. But now with government subsidy of medicine looming so large, the question of how much caring he should get inevitably becomes public.

By what standard should this public question be decided? One objective criterion—equality of access—inevitably stands out among the rest. For if we don't quite know what caring is or how much of it there should be, we can always say that at least it should be equally distributed. Medicaid has just about equalized the number of doctor visits per year between the poor and the rich. In fact, the upper class is showing a decrease in visits, and the life expectancy of richer males is going down somewhat. Presumably, no one is suggesting remedial action in favor of rich men. Equality, not health, is the issue.

EQUALITY

One can always assert that even if the results of medical treatment are illusory, the poor are entitled to their share. This looks like a powerful argument, but it neglects the Axiom of Inequality. That axiom states that every move to increase equality in one dimension necessarily decreases it in another. Consider space. The United States has unequal rates of development. Different geographic areas vary considerably in such matters as income, custom, and expectation. Establishing a uniform national policy disregards these differences; allowing local variation means that some areas are more unequal than others. Think of time. People not only have unequal incomes, they also differ in the amount of time they are prepared to devote to medical care. In equalizing the effects of money on medical care—by removing money as a consideration—care is likely to be allocated by the distribution of available time. To the extent that the pursuit of money takes time, people with a monetary advantage will have a temporal disadvantage. You can't have it both ways, as the Axiom of Allocation makes abundantly clear.

"No system of care in the world," says David Mechanic, summing up the Axiom of Allocation, "is willing to provide as much care as people will use, and all such systems develop mechanisms that ration . . . services"[83]. Just as there is no free lunch, so there is no free medicine. Rationing can be done by time (waiting lists, lines), by distance (people farther from facilities use them less than those who are closer), by complexity (forms, repeated visits, communications difficulties), by space (limiting the number of hospital beds and available doctors), or by any or all of these methods in combination. But why do people want more medical service than any system is willing to provide? The answer has to do with uncertainty.

If medicine is only partially and imperfectly related to health, it follows that doctor and patient both will often be uncertain as to what is wrong or what to do about it. Otherwise—if medicine were perfectly related to

health—either there would be no health problem or it would be a very different one. Health rates would be on one side and health resources on the other; costs and benefits could be neatly compared. But they can't because we often don't know how to produce the desired benefits. Uncertainty exists because medicine is a quasiscience—more science than, say, political science; less science than physics. How the participants in the medical system resolve their uncertainties matters a great deal.

The Medical Uncertainty Principle states that there is always one more thing that might be done—another consultation, a new drug, a different treatment. Uncertainty is resolved by doing more: the patient asks for more, the doctor orders more. The patient's simple rule for resolving uncertainty is to seek care up to the level of his insurance. If everyone uses all the care he can, total costs will rise; but the individual has so little control over the total that he does not appreciate the connection between his individual choice and the collective result. A corresponding phenomenon occurs among doctors. They can resolve uncertainty by prescribing up to the level of the patient's insurance, a rule reinforced by the high cost of malpractice. Patients bringing suit do not consider the relationship between their own success and higher medical costs for everyone. The patient is anxious, the doctor insecure; this combination is unbeatable until the irresistible force meets the immovable object—the Medical Identity.

The Medical Identity states that use is limited by availability. Only so much can be gotten out of so much. Thus, if medical uncertainty suggests that existing services will be used, Identity reminds us to add the words "up to the available supply." That supply is primarily doctors, who advise on the kind of care to provide and the number of hospital beds to maintain. But patients, considering only their own desires in time of need, want to maximize supply, a phenomenon that follows inexorably from the Principle of Perspective.

That principle states that social conditions and individual feelings are not the same thing. A happy social statistic may obscure a sad personal situation. A statistical equilibrium may hide a family crisis. Morbidity and mortality, in tabulating aggregate rates of disease and death, describe you and me but do not touch us. We do not think of ourselves as "rates." Our chances may be better or worse than the aggregate. To say that doctors are not wholly (or even largely) successful in alleviating certain symptoms is not to say that they don't help some people and that one of those people won't be me. Taking the chance that it will be me often seems to make sense, even if there is reason to believe that most people can't be helped and that some may actually be harmed. Most people, told that the same funds spent on other purposes may increase social benefits, will put their personal needs first. This is why expenditures on medical care are always larger than any estimate of the social benefit received. Now we can understand, by combining into one law the previous principles and Medical Identity, why costs rise so far and so fast.

The Law of Medical Money states that medical costs rise to equal the sum of all private insurance and government subsidy. This occurs because no one knows how much medical care ought to cost. The patient is not sure he is getting all he should, and the doctor does not want to be faulted for doing less than he might. Consider the triangular relationship between doctor, patient, and hospital. With private insurance, the doctor can use the hospital resources that are covered by the insurance while holding down his patient's own expenditures. With public subsidies, the doctor may charge his highest usual fee, abandon charitable work, and ignore the financial benefits of eliminating defaults on payments. His income rises. His patient doesn't have to pay, and his hospital expands. The patient, if he is covered by a government program or private insurance (as about 90 percent are) finds that his out-of-pocket expenses have remained the same. His insurance costs more, but either it comes out of his paycheck, looking like a fixed expense, or it is taken off his income tax as a deduction. Hospitals work on a cost-plus basis. They offer the latest and the best, thus pleasing both doctor and patient. They pay their help better; or, rather, they get others to pay their help. It's on the house—or at least on the insurance.

Perhaps our triangle ought to be a square: maybe we should include insurance companies. Why are they left out of almost all discussions of this sort? Why don't they play a cost-cutting role in medical care as they do in other industries? After all, the less the outlay, the more income for the company. Here the simplest explanation seems the best: insurance companies make no difference because they are no different from the rest of the health care industry. The largest, Blue Cross and Blue Shield, are run by the hospital establishment on behalf of doctors. After all, hospitals do not so much have patients as they have doctors who have patients. Doctors run hospitals, not the other way around. Insurance companies not willing to play this game have left the field.

What process ultimately limits medical costs? If the Law of Medical Money predicts that costs will increase to the level of available funds, then that level must be limited to keep costs down. Insurance may stop increasing when out-of-pocket payments exceed the growth in the standard of living; at that point individuals may not be willing to buy more. Subsidy may hold steady when government wants to spend more on other things or when it wants to keep its total tax take down. Costs will be limited when either individuals or governments reduce the amount they put into medicine.

No doubt the Law of Medical Money is crude, even rude. No doubt it ignores individual instances of self-sacrifice. But it has the virtue of being a powerful and parsimonious predictor. Costs have risen (and are continuing to rise) to the level of insurance and subsidy.

WHY THERE IS A CRISIS

If more than three-quarters of the population are satisfied with their medical care, why is there a crisis? Surveys on this subject are inadequate, but invariably they reveal two things: 1) the vast majority are satisfied, but 2) they wish medical care didn't cost so much and they would like to be assured of contact with their own doctor. So far as the people are concerned, then, the basic problems are cost and access. Why, to begin at the end, aren't doctors where patients want them to be?

To talk about physicians being maldistributed is to turn the truth upside down: it is the potential patients who are maldistributed. For doctors to be in the wrong place, they would have to be where people aren't, and yet they are accused of sticking to the main population centers. If distant places with little crowding and less pollution, far away from the curses of civilization, attracted the same people who advocate their virtues, doctors would live there, too. Obviously, they prefer the amenities of metropolitan areas. Are they wrong to live where they want to live? Or are the rural and remote wrong to demand that others come where they are?

Doctors can be offered a government sub-

sidy—more money, better facilities—on the grounds that it is a national policy for medical care to be available wherever citizens choose to live. Virtually all students in medical schools are heavily subsidized, so it would not be entirely unjust to demand that they serve several years in places not of their own choosing. The reason such policies do not work well—from Russia to the "Ruritanias" of this world—is that people who are forced to live in places they don't like make endless efforts to escape.

Because the distribution of physicians is determined by rational choice—doctors locate where their psychic as well as economic income is highest—there is no need for special laws to explain what happens. But the political pathology of health policy—the more the government spends on medicine, the less credit it gets—does not require explanation.

The syndrome of "the more, the less" has to be looked at as it developed over time. First we passed Medicare for the elderly and Medicaid for the poor. The idea was to get more people into the mainstream of good medical care. Following the Law of Medical Money, however, the immediate effect was to increase costs, not merely for the poor and elderly but for all the groups in between. You can't simply add the costs of the new coverage to the costs of the old; you have to multiply them both by higher figures up to the limits of the joint coverage. This is where the Axiom of Inequality takes over. The wealthier aged, who can afford to pay, receive not merely the same benefits as the aged poor, but even more, because they are better able to negotiate the system. Class tells. Inequalities are immediately created within the same category. Worse still is the "notch effect" under Medicaid, through which those just above the eligibles in income may be worse off than those below. Whatever the cutoff point, there must always be a "near poor" who are made more unequal. And so is everybody else who pays twice, first in taxes to support care for others and again in increased costs for themselves. Moreover, with increased utilization of medicine, the system becomes crowded; medical care is not only more costly but harder to get. So there we have the Paradox of Time—as things get better, they get worse.

The politics of medical care becomes a minus-sum game in which every institutional player leaves the table poorer than when he sat down. In the beginning, the number of new patients grows arithmetically while costs rise geometrically. The immediate crisis is cost. Medicaid throws state and federal budgets out of whack. The talk is all about chiselers, profiteers, and reductions. Forms and obstacles multiply. The Medical Identity is put in place. Uncle Sam becomes Uncle Scrooge. One would hardly gather that billions more are actually being spent on medicine for the poor. But the federal government is not the only participant who is doing better and feeling worse.

Unequal levels of development within states pit one location against another. A level of benefits adequate for New York City would result in coverage of half or more of the population in upstate areas as well as nearly all of Alaska's Eskimos and Arizona's Indians. The rich pay more; the poor get hassled. Patients are urged to take more of their medicine only to discover they are targets of restrictive practices. They are expected to pay deductibles before seeing a doctor and to contribute a co-payment (part of the cost) afterward. Black doctors are criticized if their practice consists predominantly of white patients, but they are held up to scorn if they increase their income by treating large numbers of the poor and aged in the ghettos. Doctors are urged to provide more patients with better medicine, and then

they are criticized for making more money. The Principle of Perspective leads each patient to want the best for himself, disregarding the social cost; and, at the same time, doctors are criticized for giving high-cost care to people who want it. The same holds true for hospitals: keeping wages down is exploitation of workers; raising them is taking advantage of insurance. Vast financial incentives are offered to encourage the establishment of nursing homes to serve the aged, and the operators are then condemned for taking advantage of the opportunity.

Does anyone win? Just try to abolish Medicare and Medicaid. Crimes against the poor and aged would be the least of the accusations. Few argue that the country would be better off without these programs than with them. Yet, as the programs operate, the smoke they generate is so dense that their supporters are hard to find.

By now it should be clear how growing proportions of people in need of medicine can be getting it in the midst of what is universally decried as a crisis in health care. Governments face phenomenal increases in cost. Administrators alternately fear charges of incompetence for failing to restrain real financial abuse and charges of niggardliness toward the needy. Patients are worried about higher costs, especially as serious or prolonged illnesses threaten them with financial catastrophe. That proportionally few people suffer this way does not decrease the concern, because it **can** happen to anyone. Doctors fear federal control, because efforts to lower costs lead to more stringent regulations. The proliferation of forms makes them feel like bureaucrats; the profusion of review committees threatens to keep them permanently on trial. New complaints increase faster than old ones can be remedied. Specialists in public health sing their ancient songs—you are what you eat, as old as you feel, as good as the air you breathe—with more conviction and less effect. True but trite: what can be done isn't worth doing, what is worth doing can't be done. The watchwords are malaise, stasis, crisis.

If money is a barrier to medicine, the system is discriminatory. If money is no barrier, the system gets overcrowded. If everyone is insured, costs rise to the level of the insurance. If many remain underinsured, their income drops to the level of whatever medical disaster befalls them. Inability to break out of this bind has made the politics of health policy pathological.

POLITICAL PATHOLOGY

Health policy began with a laudable effort to help people by resolving the polarized conflict between supporters of universal, national health insurance ("socialized" medicine) and the proponents of private medicine. Neither side believed a word uttered by the other. The issue was sidestepped by successfully implementing medical care for the aged under Social Security. Agreement that the aged needed help was easier to achieve than consensus on any overall medical system. The obvious defect was that the poor, who needed financial help the most, were left out unless they were also old and covered by Social Security. The next move, therefore, was Medicaid for the poor, at least for those reached by state programs.

Even if one still believed that medicine equaled health, it became impossible to ignore the evidence that availability of medical services was not the same as their delivery and use. Seeing a doctor was not the same as actually doing what he prescribed. It is hard to alleviate stress in the doctor's office when the patient goes back to the same stress at home and on the street.

"Health delivery" became the catchword. At times it almost seemed as if the welcome wagon was supposed to roll up to the door and deliver health, wrapped in a neat package. One approach brought services to the poor through neighborhood health centers. The idea was that local control would increase sensitivity to the patients' needs. But experience showed that this "sensitivity" had its price. Local "needs" encompassed a wider range of services, including employment. The costs per patient-visit for seeing a doctor or social worker were three to four times those for seeing a private practitioner. Achieving local control meant control by inside laymen rather than outside professionals, a condition doctors were loath to accept. Innovation both in medical practice and in power relationships proved a greater burden than distant federal sponsors could bear, so they tried to coopt the medical powers by getting them to sponsor health centers. The price was paid in higher costs and lower local control. Amid universal complaints, programs were maintained where feasible, phased out where necessary, and forgotten where possible.

By now the elite participants have exceeded their thresholds of pain: government can't make good on its promises to deliver services; administrators are blamed for everything from malpractice by doctors to overcharges by hospitals; doctors find their professional prerogatives invaded by local activists from below and by state and federal bureaucrats from above. From the left come charges that the system is biased against the poor because local residents are unable to obtain, or maintain, control of medical facilities, and because the rates by which health is measured are worse for them than for the better off. Loss of health is tied to lack of power. From the right come charges that the system penalizes the professional and the productive: excessive governmental intervention leads to lower medical standards and higher costs of bureaucracy, so that costs go up while health does not.

As neighborhood health centers (NHCs) phased out, the new favorites, the health maintenance organizations (HMOs), phased in. If the idea behind the NHCs was to bring services to the people, the idea behind the HMOs is to bring the people to the services. If a rationale for NHCs was to exert lay control over doctors, the rationale for HMOs is to exert medical control over costs. The concept is ancient. Doctors gather together in a group facility. Individuals or groups, such as unions and universities, join the HMO at a fixed rate for specified services. Through efficiencies in the division of labor and through features such as bonuses to doctors for less utilization, downward control is exerted on costs.

Since the basic method of cutting costs is to reduce the supply of hospital beds and physician services (the Medical Identity), HMOs work by making people wait. Since physicians are on salary, they must be given a quota of patients or a cost objective against which to judge their efforts. Both incentives may have adverse effects on patients. HMO patients complain about the difficulty of building up a personal relationship with a doctor who can be seen quickly when the need arises. Establishing such a relationship requires communication skills most likely to be found among the middle class. The patient's ability to shop around for different opinions is minimized, unless he is willing to pay extra by going outside the system. Doctors are motivated to engage in preventive practices, though evidence on the efficacy of these practices is hard to come by. They are also motivated to engage in bureaucratic routines to minimize the patients' demands on their time; and they may divert patients to various specialties or ask them to return, so as to fit them into each physician's assigned quota. In a word, HMOs are a mixed bag, with no one quite sure yet what the trade-off is between effi-

ciency and effectiveness. Turning the Great Equation into an Identity— where Health = Health Maintenance Organization—does, however, solve a lot of problems by definition.

HMOs may be hailed by some as an answer to the problem of medical information. How is the patient-consumer to know whether he is getting proper care at reasonable cost? If it were possible to rate HMOs, and if they were in competition, people might find it easier to choose among them than among myriad private doctors. Instead of being required to know whether all those tests and special consultations were necessary, or how much an operation should cost, the patients (or better still, their sponsoring organizations) might compare records of each HMO's ability to judge. Our measures of medical quality and cost, however, are still primitive. Treatment standards are notoriously subjective. Health rates are so tenuously connected to medicine that they are bound to be similar among similar populations so long as everyone has even limited access to care.

If health is only minimally related to care, less expertise may be about as good as more professional training. If by "care" many or most people mean simply a sympathetic listener as much as, or more than, they mean a highly trained, cold diagnostician, cheaper help may be as good as, or even better than, expensive assistance. Enter the nurse-practitioner or the medical corpsman or the old Russian **feldsher**—medical assistants trained to deal with emergencies, make simple diagnoses, and refer more complicated problems to medical doctors. They cost less, and they actually make home visits. The main disadvantage is their apparent challenge to the prestige of doctors, but it could work the other way around: doctors might be elevated because they deal with more complicated matters. But the success of the medical assistant might nonetheless raise questions about the mystique of medical doctors. In response the doctors might deny that anyone else can really know what is going on and what needs to be done, and they might then use assistants as additions to (but not substitutes for) their services. That would mean another input into the medical system and therefore an additional cost. The politics of medicine is just as much about the power of doctors as it is about the authority of politicians.

Now we see again, but from a different angle, why the medical system seems in crisis although most people are satisfied with the care they are receiving. At any one time, most people are reasonably healthy. When they do need help, they can get it. The quality of care is generally impressive; or whatever ails them goes away of its own accord. But these comments apply only to the mass of patients. The elite participants—doctors, administrators, politicians—are all frustrated. Anything they turn to rebounds against them. Damned if they do and cursed if they don't, it is not surprising that they feel that any future position is bound to be less uncomfortable than the one they hold today. Things can always get worse, of course, but it is not easy for them to see that.

GOVERNMENTAL LEGITIMACY: CURING THE SICKNESS OF HEALTH

Why should government pay billions for health and get back not even token tribute? If government is going to be accused of abusing the poor, neglecting the middle classes, and milking the rich; if it is to be condemned for bureaucratizing the patient and coercing the doctor, it can manage all that without spending billions. Slanders and calumnies are easier to bear when they are cost-free. Spending more for worse treatment is as bad a policy for government as it

would be for any of us. The only defendant without counsel is the government. What should it do?

The Axiom of Inequality cannot be changed; it is built into the nature of things. What government can do is to choose the kinds of inequalities with which it is prepared to live. Increasing the waiting time of the rich, for instance—that is, having them wait as long as everybody else—may not seem outrageous. Decreasing subsidies in New York City and increasing them in Jacksonville may seem a reasonable price to pay for national uniformity. From the standpoint of government, however, the political problem is not to achieve equal treatment but to get support, at least from those it intends to benefit. Government needs gratitude, not ingrates.

The Principle of Goal Displacement, through the double-displacement effect, succeeds only in substituting access to care for health; it by no means guarantees that people will value the access they get. Equal access to care will not necessarily be equated with the best care available or with all that patients believe they require. Government's task is to resolve the Paradox of Time so that, as things get better, people will see themselves as better off.

Proposals for governmental support of medical care have ranged from modest subsidies to private insurance (the AMA's Medicredit) to public control of the medical industry on the British model. The latter has never had much support in this country, because of the widespread opposition to socializing doctors by turning them into de facto government employees. The former has lost whatever support it once had as respect for the AMA has declined, its internal unity has diminished, and its congressional supporters have nearly vanished. Private insurance seems as much the problem as the solution.

The two most prominent proposals would resolve the political problems of medical care in contrasting ways, but substantively they are similar. Both the Comprehensive Health Insurance Plan (CHIP), introduced in the last days of the Nixon administration, and the Kennedy-Mills proposal would involve billions of dollars in additional expenditures. Estimates put each of them at $42 billion to start, less substantial existing expenditures—but then no estimates in this field have ever come remotely close to reality. Both proposals would provide health insurance for virtually everyone and would cover almost everything (including catastrophic and long-term illness) except for prolonged mental illness and nursing-home care. Both include a string of deductibles and coinsurance mechanisms, with CHIP so complex as almost to defy description. Both seek to hold down costs by giving individuals a financial incentive to limit use. Neither provides incentives for the medical community to contain costs, other than the importunings of insurance companies and state governments (CHIP) or the federal government (Kennedy-Mills), which have not been noticeably effective in the past.

CHIP would be financed largely through employer-employee contributions, with employees making a per capita payment; Kennedy-Mills substitutes a more (though by no means entirely) progressive proportionate tax. CHIP mandates insurance and gives a choice of private plans supervised through state agencies. Kennedy-Mills works largely through a special fund collected and administered by the federal government. The basic difference between them is that more of the cost of Kennedy-Mills shows up in the federal budget, while most of the cost of CHIP, as its acronym suggests, is diffused through the private sector.

The most likely consequence of both proposals would be a vast inflation of costs without a corresponding increase in services. Since medical manpower and facilities could not increase proportionately with demand, prices would rise. It would be Medicaid all over again, only worse because so many new things would be attempted and so many old things expanded. Almost before the ink dried on the legislation, efforts would be under way to delay this provision, lessen the cost of that one, introduce rationing in nonmonetary ways, find more forms for doctors and patients to fill out, and on and on. Cries of systemic crisis would be replaced by prophecies of systemic failure. But enough. My purpose is not to predict the medical consequences of these proposals but to analyze their political rationale.

Based on the political premise that some form of national health insurance was inevitable, CHIP sought to limit the government's liability. By joining the opposition, the Nixon administration hoped to control the apparatus so as to lessen its impact on the federal budget and bureaucracy. If people were determined to have something that wasn't going to help them, the government could at least see to it that the totals did not swamp its budget or overload its administration. The costs of failure would be spread around among the states, the various insurance companies, and innumerable individual and group medical practices. Just as revenue sharing was designed to channel demands to state and local governments, instead of the national government (here's a little money and a lot of trouble, and don't bother me!) so CHIP was devised to diffuse responsibility.

What the Republican administration did not foresee was that the rapid breakdown of the existing medical system would inevitably lead to demands for a federal takeover. When a company goes bankrupt, it is usually returned, not to its owners, but to its creditors. This insight belongs to the sponsors of the Kennedy-Mills bill. They seized on the Nixon plan to advance one that would load additional clients, services, and billions onto the shoulders of government. Wouldn't this proposal be too expensive and cumbersome? The worse the better, politically! For then the stage would be set for a national health service.

Under the Kennedy-Griffith (now Kennedy-Corman) proposal, which was the senator's original preference, every person in the United States would, without personal payment, be covered for a wide variety of services, thus replacing all public programs and private insurance with an all-inclusive federal system. Every public and personal medical expense would be transferred to the federal government, paid for half by additional payroll taxes and higher taxes on unearned income and half from general revenues. Obviously, as the sole direct payer, the federal government would have control over costs, but, by the same token, it would have to make all the decisions on how much of what service would be provided to which people in what way for how long.

The difference between Kennedy-Mills and the Kennedy-Corman Health Security Act (HSA) is that the latter would work directly on the Law of Medical Money by limiting the financial resources flowing into the medical system. Whatever the federal government allocated would be all that could be spent, except for the sums spent by those people choosing to pay extra to go outside the system. To put HSA in proper perspective, it is useful to contrast it with another proposal, one that would also limit supply but from a different direction. Senators Long and Ribicoff proposed to deal with the costs of catastrophic illness by setting individual-expenditure limits beyond which costs would be paid by the government. But Long-Ribicoff did not relate individual payments to income. For our comparison, therefore, it

is more helpful to concentrate on Martin Feldstein's proposal for an income-graded program in which each person pays medical costs up to a specified proportion of his income, after which the government picks up the remaining (defined as catastrophic) expenses. Medicare and Medicaid are replaced, as all benefits are related to income. The poor pay less, the rich pay more, but everyone is protected against the costs of catastrophe. Although the catastrophic portion would rise in cost, especially for long-term disability, it would represent a relatively small proportion of medical expenditures. Total costs would be determined by overall financial inputs, which would be limited by the willingness of people to pay instead of inflated by using up their insurance or subsidy.

At first glance it might appear strange for national health insurance (whether through private intermediaries or direct government operation) to be conceived of as a method for limiting costs; but experience in practice, as well as deduction from theory, bears out that conception. The usual complaint in Britain, for example, is that the National Health Service is being starved for funds: hospital construction has been virtually nil; the number of doctors per capita has hardly increased; long queues persist for hospitalization in all but emergency cases. Why? Because health care accounts for a sizable proportion of both government expenditure and gross national product and must compete with family allowances, housing, transportation, and all the rest. While there are pressures to increase medical expenditures, they are counterbalanced by demands from other sectors. In times of extreme financial stringency, all too frequent as government expenditure approaches half of the GNP, it is not likely that priority will go to medicine.

So much for current trends. In the future, the nation will probably move toward (and vacillate between) three generic types of health-care policies: 1) a mixed public and private system like the one we have now, only bigger; 2) total coverage through a national health service; and 3) income-graded catastrophic health insurance. It will be convenient to refer to these approaches as "mixed," "total," and "income."

The total and income approaches have weaknesses. The income-catastrophic approach might encourage a "sky's the limit" attitude toward large expenditures; the other side of the coin is that resources would flow to those chronically and/or extremely ill people who most need help. The total approach would strain the national budget, putting medical needs at the mercy of other concerns, such as tax increases; on the other hand, making medicine more political might have the advantage of providing more informed judgment on its relative priority. The two approaches, however, are more interesting for their different strengths than for their weaknesses.

The income approach would magnify individual choice until the level of catastrophic cost is reached. Holding ability to pay relatively constant, each person would be able to decide how much (in terms of what money can buy) he is willing to give up to purchase medical services. There would be no need to regulate the medical industry as to cost and service: supply and demand would determine the price. Paperwork would be minimized. So would bureaucracy. Under- or over-utilization could be dealt with by raising or lowering the percentage limits at each level of income, rather than by dealing with tens of thousands of doctors, hospitals, pharmacies, and the like. The total approach, by contrast, could promise a kind of collective rationality in the sense that the government would make a more direct determination of how much the nation wanted to spend on health versus other desired expenditures.

How might we choose between an essentially administrative and a primarily market-oriented mechanism? Each is as political as the other, but they come to their politics in different ways. An income approach would be simpler to administer and easier to abandon. If it didn't work, more ambitious programs could readily be subsidized. A total approach could promise more, because no one under existing programs would be worse off (except taxpayers), and everyone with insufficent coverage would come under its comprehensive umbrella. The backers of totality fear that the income approach would preempt the health field for years to come. The proponents of income grading fear that, once a comprehensive program is begun, there will be no getting out of it—too many people would lose benefits they already have, and the medical system would have unalterably changed its character. The choice (not only now but in the future) really has to be made on fundamental grounds of a modified-market versus an almost entirely administrative approach. Which proposal would be not only proper for the people but good for the government?

MARKET VERSUS ADMINISTRATIVE MECHANISMS

At the outset, I should state my conviction that doing either one consistently would be better than mixing them up. Both methods would give government a better chance to know what it is doing and to get credit for what it does. Expenditures on the medical system, whether too high or too low for some tastes, would be subject to overall control instead of sudden and unpredictable increases. Patients would have a system they could understand and would therefore be able to hold government accountable for how it was working. Under one system they would know that care was comprehensive, crediting government with the program and criticizing it for quality and cost. Under the other, they would know they were being encouraged to exercise discretion, but within boundaries guaranteeing them protection against catastrophe. Under the present system, they can't figure out what's going on (who can?); or why their coverage is inadequate; or why, if there is no effective government control, there are so many governmental forms. Mixed approaches will only exacerbate these unfortunate tendencies, multiplying ambiguities about deductibles and co-payment amid startling increases in cost. If we want our future to be better than our past, then let us look more closely at the bureaucratic and market models for medical care.

What do we know about medical care in a bureaucratic setting? Distressingly little. But there may be just enough collected from studies of HMOs and of systems in other countries, especially Britain, to provide a few clues. Doctors in HMOs work fewer hours than do doctors in private practice. This is not surprising. One of the attractions of HMOs for doctors is the limit on the hours they can be put on call. Market physicians respond to increases in patient load by increasing the hours they see patients; physicians working in a bureaucratic context respond by spending less time with patients. Two consequences of a public system are immediately apparent: more doctors will be needed, and less time will be spent listening and examining. Patients' demands for more time with the doctor will be met by repeated visits rather than longer ones. But will doctors be distributed more equally over the nation? The evidence suggests not. Britain has failed to achieve this goal in the quarter-century since the National Health Service began. The reason is that not only economic but also political allocations are subject to biases, one of which, incidentally, is called

majority rule. The same forces that gather doctors in certain areas are reflected in the political power necessary to supply funds to keep them there.

Surely the ratio of specialists to general practitioners could be better controlled by central direction than by centrifugal market forces. Agreed. But a price is paid that should be recognized. The much higher proportion of general practitioners in Britain is achieved through a class bias that values "consultants" (their "specialists") more highly than ordinary doctors. (Consultants are called "Mister," as if to emphasize their individual excellence, while general practitioners are given the collective title of "Doctor.") The much higher proportion of specialists in America may stem in part from a desire to maintain equality among doctors—a nice illustration of the Axiom of Inequality. One result of the British custom is to lower the quality of general practice; another is to deny general practitioners access to hospitals. They lose control of their patients at the portal, leaving them without the comfort they may need in a stressful time and subjecting them to a bewildering maze of specialists and subspecialists, separated by custom and procedure, none of whom may be in charge of the whole person.

Would a bureaucratic system based on fixed charges and predetermined salaries place more emphasis on cheaper prevention than on more expensive maintenance, or on outpatient rather than hospital service? Possibly. (No one knows for sure whether preventive medicine actually works.) Doctors, in any event, do not cease to be doctors once they start operating in a bureaucratic setting. Cure, to doctors, is intrinsically more interesting than prevention; it is also something they know they can attempt, whereas they cannot enforce measures such as "no smoking." If it were true, moreover, that providing ample opportunities to see doctors outside the hospital would reduce the need to use hospitals, then providing outpatient services should hold down costs. The little evidence available, however, suggests otherwise. A natural experiment for this purpose takes place when patients have generous coverage for both in- and outpatient medical services. Visits to the doctor go up, but so does utilization of hospitals. More frequent visits generate awareness of more things wrong, for which more hospitalization is indicated. The way to limit costs, if that is the objective, is to limit access to hospitals by reducing the number of available beds.

The great advantage of a comprehensive health service is that it keeps expenditures in line with other objectives. The Principle of Perspective works both ways: if an individual is not an aggregate, neither is an aggregate an individual. Left to our own devices, at near zero cost, you and I use as much as we and ours need. At the government level, however, it is not a question of personal needs and desires but of collective choice among different levels of taxation and expenditure. Hence, it should not be surprising that our collective choice would be less than the summed total of our individual preferences.

The usual complaint about the market method is money. Poor people are kept out of the medical system by not having enough. No one disputes this. And whatever evidence exists also suggests that the use of deductibles and co-payment exerts a disproportionate effect in deterring the poor from acquiring medical care. Therefore, to preserve as much of the market as possible, the response is to provide the poor with additional funds they can use for any purpose they desire, including (but not limited to) medical care. This immediately raises the issue of services in kind versus payment in cash. Enabling the poor to receive medical services

without financial cost to themselves means they cannot choose alternative expenditures. A negative way of looking at this is to say that it reveals distrust of the poor: presumably, the poor are not able to make rational decisions for themselves, so the government must decide for them. A positive approach is to say that health is so important that society has an interest in assuring that the poor receive access to care. I almost said, "whether they want it or not," but, the argument continues, the choice of seeking or not seeking health care is neither easy nor simple: the poor—because they are poor, because money means more to them, because they have so many other vital needs—are under great temptation to sacrifice future health to present concerns. The alleged short-sighted psychology of the poor requires that they be protected against themselves.

The problem is not with the intellectually insubstantial (though politically potent) arguments that medical care is a right and that money should have nothing to do with medicine. The Axiom of Allocation assures us that medical care must be allocated in some way, and that, if it is not done at the bottom through individual income, it will be done at the top through national income. If medicine is a right, so is education, housing, food, employment (without which other rights can no longer be enjoyed), and so on until we are led to the same old problems of resource allocation. The real question is whether care will be allocated by governmental mechanisms, in which one-man, one-vote is the ideal, or by the distribution of income, in which one-dollar, one-preference is the ideal, modified to assist the poor.

The problem for market men is not to demonstrate resource scarcity but to show that one of the essential conditions of buying and selling really is operating. I refer to consumer information about the cost and quality of care. The same problems crop up in many other areas involving technical advice: without knowing as much as the lawyer, builder, garage mechanic, or television repair man, how can the consumer determine whether the advice is good and the work performed properly and at reasonable cost?

The image in the literature is amateur patient versus professional doctor: the patient is not sure what is wrong, who the best doctor is, and how much the treatment should cost. Worse still, doctors deliberately withhold information by making it unethical to advertise prices or criticize peers. Should the doctor be less than competent or more than usually inclined to run up a bill, there is little the patient can do.

There are elements of reality in this picture, as all of us will recognize, but it is exaggerated. People can and do ask others about their experiences with various doctors: mothers endlessly compare pediatricians, for example. The abuses with which we are concerned are more likely to occur when patients lack a stable relationship with at least one doctor, and when there is no community whose opinions the doctor values and the patient learns to consult.

Nevertheless, it is obvious that patient-consumers do lack full information about the medical services they are buying. So, in fact, do doctors lack full knowledge of the services they are selling. How, then, might the imperfect medical market be improved? Would some alternative provision of medical services ensure better information?

Since all costs would be paid by taxpayers, government would have an incentive to keep the expenditures on a national health service in proportion to the expenditures for other vital activities. The very feature that has so far made a national health service politically unpalatable—it would take over about $50 billion of now private expenditures, thus requiring a massive tax increase—would im-

mediately make the government financially responsible. Under a total governmental program, central authorities would have to determine how much should be spent and how these funds should be allocated to regional authorities. Basing the formula on numbers would put remote places at a disadvantage; basing it on area would put populous places at a disadvantage. How would regional authorities decide how much money to put toward hospital beds versus outpatient clinics, versus drugs, versus long-term care? There are few objective criteria. Would teams of medical specialists make the decisions? Professional boundaries would cause problems. Would administrators? Lack of medical expertise would cause problems. Administrative committees would have to decide who receives how much treatment, given the limited resources available from the central authority. Would their collective judgment be better or worse than that of individuals negotiating with doctors and hospitals? No one knows. But something can be said about the trade-off between quality and cost.

Suppose the question is: Under which type of system are costs likely to be highest per capita? The answer is: first, mixed public and private; second, mostly private; third, mostly public. Costs are greater under a mixed system because potential quality is valued over real cost: it pays each individual to use up his insurance and subsidy, because the quality/cost ratio is set high. Under the mostly private system, the individual has an incentive to keep his costs down. Under the largely public one, the government has an incentive to keep its costs within bounds. Because each individual regards his personal worth more than his social value, however, a series of individual payments will add up to something more than the payments determined by the very same people's collective judgment. At the margins, then, the economic market, preferring quality over cost, would produce somewhat larger expenditures than would the political arena.

Who would value a public medical system? Those who want government to exert maximum control over at least cost. The term "cost" here may be used in two ways—financial and political. Government does more, is able to allocate more resources, and has more of a chance of getting support for what it does. People who are more concerned with equality than with quality of care—though, of course, they want both—also should prefer public financing. It assures reasonably equal access, and it also places medical care in the context of other public needs. Doctors who value independence and patients who value responsiveness would be less in favor of a public system.

Who would prefer a private system, providing the effects of income were mitigated? People who want less governmental direction and more personal control over costs. These include doctors who want less governmental control, patients who want more choice, and politicians who want more leeway in resource allocation and less blame for bureaucratizing medicine.

I would prefer the income approach, because it is readily reversible; it means less bureaucracy and more choice. The total approach, however, could be infused with choice: Under the rubric of a single national health service, there could be three to six competitive and alternative programs, each organized on a different basis. There could be HMOs, foundation plans (under which individual doctors contract with a central service), and other variants. Patients could use any of these programs, all of which would be competing for their favor. The total sum to be spent each year would be fixed at the federal level, and each service would be paid its proportionate share ac-

cording to the number and type of patients it had enrolled. Thus, we could mitigate the worst features of a bureaucratic system while maintaining its strengths.

THOUGHT AND ACTION

Let us summarize. Basically there are two sites for relating cost to quality—that is, for disciplining needs, which may be infinite, by controlling resources, which are limited. One is at the level of the individual; the other, at the level of the collectivity. By comparing his individual desires with his personal resources, through the private market, the individual internalizes an informal cost-effectiveness analysis. Since incomes differ, the break-even point differs among individuals. And if incomes were made more equal, individuals would still differ in the degree to which they choose medical care over other goods and services. These other valued objects would compete with medicine, leading some individuals to choose lower levels of medicine and thus reducing the inputs into (and cost of) the system. This creative tension can also be had at the collective level. There it is a tension between some public services, such as medicine, and others, such as welfare, and a tension between the resources left in private hands and those devoted to the public sector. The fatal defect of the mixed system, a defect that undermines the worth of its otherwise valuable pluralism, is that it does not impose sufficient discipline either at the individual or at the collective level. The individual need not face his full costs, and the government need not carry the full burden.

My purpose in writing this essay has not been to assess current political feasibility but to determine longer-lasting political virtue. The proposals I believe to be the worst for sustaining the legitimacy of government are at present the most popular. Proposals that deserve the most serious attention are ignored. The falsely assumed excessive cost of total care and the falsely believed inequality of the income approach have removed them from serious consideration. Perhaps this is the way it has to be. But I believe there is still time to change our ways of thinking about the medical care. Medicine is by no means the only field where how we think affects what we believe, where what we believe is the key to how we feel, and where how we feel determines how we act.

If politicians did not believe that better health would emerge from greater effort, could they justify pouring billions more into the medical system? It could be argued that belief in medicine—doctor as witchdoctor—is so deeply ingrained that no evidence to the contrary would be accepted. Maybe. But this argument does not reach the question of what politicians would do if they believed otherwise.

Suppose the people were told that additional increments devoted to medicine would not improve their collective health but would give them more opportunity to express their individual feelings to doctors. How much more would they pay for this "caring"? Would it be as much as $10 billion? Would it be that high if the program contained no guarantee—and none do—that doctors would care more or be more available?

In any event, after the mixed approach fails, as it surely will, this country will be faced with the same alternatives—putting together the pieces administratively through a national health service, or dismantling what exists in favor of a modified market mechanism. But this is all too neat.

It could be, of course, that the future will find the worst is really the best. The three systems I have separated for analytical convenience—private, public, and mixed—may in practice refuse to reveal their pristine purity. What life has joined together no abstrac-

tion may be able to put asunder. A national health service, for instance, might quickly lose its putatively public character as numerous individuals opt for private care. In Scandinavian countries, even those in the professional strata who are convinced supporters of public medicine often prefer to use private doctors. They pay out to jump queues, so as to be treated when they wish and to have private hospital rooms to carry on business or just to receive extra attention. By paying twice, once through taxes and once through fees for service, they raise the total cost of medicine to society. Would not a public system that was 20 or 30 percent private be, in reality, mixed?

Consider an income-graded catastrophic system. It would, to begin with, have to pay all costs for those below the poverty line. As time passed, political pressure might increase the proportion of the population subsidized to 25 or 30 percent. As costs increased, administrative action might be undertaken to limit coverage of expensive long-term illness. How different, then, would this presumably private system be from the mixed system it was designed to replace?

The present as future may be replaced by the future as future only to be superseded by the future as past. First the mixed system (the present as future) will be intensified by pouring billions into it (a la Kennedy-Mills). When that fails, an income-graded catastrophic plan or a national health service (the future as future) will be tried. Efforts to make the former system wholly private will be unfeasible, because public sentiment is against rationing medical care solely by money. Efforts to make the latter system wholly public will fail, because forbidding private fees for service will appear to citizens as an intolerable restraint on their liberty. Then we can expect the future as past. By the next century, we may have learned that a mixed system is bad in every respect except one—it mirrors our ambivalence. Whether we will grow up by learning to live with faults we do not wish to do without is a subject for a seer, not a social scientist.

Health policy is pathological because we are neurotic and insist on making our government psychotic. Our neurosis consists in knowing what is required for good health (Mother was right: Eat a good breakfast! Sleep eight hours a day! Don't drink! Don't smoke! Keep clean! *And* don't worry!) but not being willing to do it. Government's ambivalence consists in paying coming and going: once for telling people how to be healthy and once for paying their bills when they disregard this advice. Psychosis appears when government persists in repeating this self-defeating play. Maybe twenty-first-century man will come to cherish his absurdities[85].

REFERENCES AND NOTES

1. Aldrich, H. Organizational boundaries and interorganizational conflict. *Human Relations,* 24:279-93, 1971.
2. Buckley, W. *Sociology and Modern Systems Theory.* Englewood Cliffs, N.J.: Prentice-Hall, 1967, p. 58.
3. Miller, E., and Rice, A. *Systems of Organization: The Control of Task and Sentient Boundaries.* London: Tavistock, 1967, p. 9.
4. Miller, E., and Rice, A. 1967, p. 266.
5. Evans, W.H. *Organization Theory: Structures, Systems, and Environments.* New York: John Wiley, 1976, p. 124.
6. Thompson, J.D. *Organizations in Action.* New York: McGraw-Hill, 1967, pp. 66-73.
7. Rushing. W., and Zald, M. (Eds.) *Organizations and Beyond: Selected Essays of James D. Thompson.* Lexington, Mass.: Lexington Books, 1976, p. ix.
8. Belknap, I. *The Human Problems of a State Mental Hospital.* New York, 1956.
9. Lentz, E.M. Hospital administration—one of a series. *Admin. Sci. Quart.,* 1:444-63, 1957.
10. Elling, R.H., and Halebsky, S. Organizational differentiation and support: a conceptual framework. *Admin. Sci. Quart.,* 6:185-209, 1961.
11. March, J.G., and Simon, H.A. *Organizations.* New York, 1958.
12. Dill, W.R. Environment as an influence on managerial autonomy. *Admin. Sci. Quart.,* 2:409-43, 1958.
13. Coleman, J.S. *Community Conflict.* New York, 1957.
14. Dill, W.R. 1958.
15. Chandler, A.D., Jr. *Strategy and Structure.* Cambridge, Mass., 1962.
16. Dinerman, H. Image Problems for American Companies Abroad. In Riley, J.W. (Ed.) *The Corporation and Its Publics.* New York, 1963. Also Chandler, A.D., Jr. 1962.
17. Blau, P.M. *The Dynamics of Bureaucracy.* Chicago, 1955.
    Francis, R.G., and Stone, R.C. *Service and Procedure in Bureaucracy.* Minneapolis, 1956.
18. Weber, M. *The Theory of Social and Economic Organization.* Henderson, A.M., and Parsons, T. (Trans.) and Parsons, T. (Ed.) New York: Free Press of Glencoe, 1947.
    Merton, R.K. The Role-set: problems in sociological theory. *Brit. J. Sociology,* 8:106-20, June 1957.
19. March, J.G., and Simon, H. *Organizations.* New York: John Wiley, 1958.
20. Thompson, J.D. 1967, p. 147.
21. Thompson, J.D. 1967, p. 158.
22. Negandhi, A. A Step beyond in the Present. In Negandhi, A. (Ed.) *Interorganization Theory.*

Kent, Ohio: Kent State University Press, 1975, p. 260.
23. Crozier, M. The Cultural Determinants of Organizational Behavior. In Negandhi, A. (Ed.) *Modern Organization Theory.* Kent, Ohio: Kent State University Press, 1973.
24. Katz, D., and Kahn, R.L. *The Social Psychology of Organizations,* 2nd ed. New York: John Wiley, 1978, p. 131.
25. Thompson, J.D. 1967.
26. Thompson, J.D., and McEwen, W. Organizational goals and environment: goal setting as an interaction process. *Am. Sociological Rev.* 23:23-31, Feb. 1958.
27. Cyert, R., and March, J.G. *A Behavioral Theory of the Firm.* Englewood Cliffs, N.J.: Prentice-Hall, 1963.
28. Macauley, S. Non-contractual relations in business: a preliminary study. *Am. Sociological Rev.,* 28:55-67, Feb. 1963.
29. Selznick, P. *TVA and the Grass Roots.* Berkeley, Calif.: University of California Press, 1949.
30. The 10-day orientation includes three days of classroom instruction on hospital and nursing procedures, safety, and other pertinent topics and seven days of working on the clinical units with only feedback, review, and counseling from the staff development department.
31. Evans. W.A. 1976, p. 284.
32. Benson, J.K. The interorganizational network as a political economy. *Admin. Sci. Quart.,* 20: 229-49, June 1975.
33. Benson, J.K. 1975, p. 280.
34. Benson, J.K. 1975, p. 280.
35. Freidson, E. *Professional Dominance: The Social Structure of Medical Care.* New York: Atherton, 1970.
36. An abbreviated version of this paper appeared in 1974 as "Health planning in Appalachia: conflict resolution in a community hospital." *Soc. Sci. Med.,* 8:135-42.
37. Alford labels this coalition of interest groups the "equal-health advocates" and shows how their efforts are likely to fail. See Alford, R.R. The political economy of health care: dynamics without change. *Politics and Society,* Winter 1972.
38. All names introduced in quotation marks are fictitious.
39. Freidson, E. 1970, pp. 209-37.
40. From a letter from the board chairman of the hospital to the medical staff, October 1945.
41. From the minutes of the board meeting.
42. "Even today a vast bulk of the total surgical work in the United States is undertaken by doctors regarded as 'non-qualified' surgeons by the American College of Surgeons." Mechanic, D. The changing structure of medical practice. *Law*

and Contemporary Problems, p. 708, Autumn, 1967.

43. Pellgrin, R.J., and Coates, C. Absentee-owned corporations and community power structure. *Am. J. Sociology,* 61:413–19, 1956.

44. McEashern, M.T. *Hospital Organization and Management.* Berwyn, Ill.: Physician Record, 1957.

45. The two hospitals decided to merge boards and staffs as of 1972.

46. Weinerman, E.R. Problems and Perspectives of Group Practice. In Elling, R.H. (Ed.) *National Health Care.* New York: Aldine, 1971, pp. 206–20.

47. Roemer, M.I., and Wilson, E.A. Organized Health Services in a County in the United States. Health Service Publication 197, 1952.

48. McNerney, W., and Riedel, D.C. Regionalization and Rural Health Care. Ann Arbor: University of Michigan Press, 1962.

49. Clark, H.T., Jr. The challenge of the regional medical programs legislation. *J. Med. Ed.,* 41:344–61, 1966.

50. Schulze, R.O. The Bifurcation of Power in a Satellite City. In Janowitz, M. (Ed.) *Community Political Systems.* Glencoe, Ill.: Free Press, 1961, pp. 19–80.

51. Presthus, R.V. The social basis of bureaucratic organization. *Social Forces,* 38:103–109, 1959.

52. Bellin, L. Changing Composition of Voluntary Hospital Boards—Inevitable Prospect for the 1970's. New York: New York Health Department, 1971.

53. Balbus, I.D. The concept of interest in pluralist and Marxian analysis. *Politics and Society,* 1:515–77, 1971.

54. Freidson, E. 1970, p. 168.

55. Bodenheimer, T.S. Regional medical programs: no road to regionalization. *Med. Care Rev.,* 26:1125–66, 1969. See also Krause, E.A. Health planning as a managerial ideology. *Int. J. Health Serv.,* 3:445–63, 1973.

56. Alford, R.R. 1972, p. 128.

57. Galbraith, J.K. *Economics and the Public Purpose.* Boston: Houghton Mifflin, 1973.

58. Gish, O. Resource allocation, equality of access, and health. *Int. J. Health Serv.,* 3:399–412, 1973.

59. Supported by a contract with the Bureau of Health Resources Development, Department of Health, Education and Welfare (1-MB-24377).

60. Supported by a Nursing Special Project Grant, Public Health Service, Department of Health, Education and Welfare (5 D10 NU 01558-03).

61. Kramer, M. *Reality Shock: Why Nurses Leave Nursing.* St. Louis: C.V. Mosby, 1974.
Smith, K. Discrepancies in the role-specific values of head nurses and nursing educators. *Nurs. Res.,* 14(3):196–200, 1965.

62. Armstrong, M.L. Bridging the gap between graduation and employment. *J. Nurs. Admin.* 4(6):42–48, 1974.

Schein, E. The first job dilemma. *Psych. Today,* 1(10):27–37, 1968.

63. All results described as "significant" have demonstrated a 0.05, 0.02, or 0.01 level of confidence on the Mann Whitney U or Wilcoxen T test; all results described as "trends" have shown at least a 0.01 level of confidence on the U or T test. (Contact the authors for complete statistical analysis.)

64. Munson, F., and Heda, S. An instrument for measuring nursing satisfaction. *Nurs. Res.,* 23(2):159–66, 1974.

65. Meleis, A. *Operation Concern: A Study of Senior Nursing Students.* Report for the Committee on Nursing and Nursing Education, The San Francisco Consortium, San Francisco, Aug. 1973.
Seeman, M., and Evans, J. Alienation and learning in a hospital setting. *Am Sociological Rev.,* 27(6):772–82, 1962.

66. Weiss, S., and Ramsey, E. *Area Health Education Center: Nurse Orientation.* Report for the Nursing Education Committee and the Nurse Administrators Council of the Central San Joaquin Valley, Calif., March 1976.

67. Benner, P., and Benner, R. *The New Graduate—Perspectives, Practice and Promise* Report of the Coordinating Council for Education in Health Sciences in San Diego and Imperial Counties, Calif., June 1975.

68. Corwin, R. The professional employee: a study of conflict in nursing roles. *Am. J. Sociology,* 66(6):604–15,1961.

69. Leavitt, H.J., Dill, W.R., and Eyring, H.B. *The Organization World.* New York: Harcourt Brace Jovanovich, 1973, p. 327.

70. Leavitt, H.J., Dill, W.R., and Eyring, H.B. 1973, p. 329.

71. Roemer, M.I. From poor beginnings: the growth of primary care. *Hospital,* pp. 38–43, March 1, 1975.

72. Thompson, J.D. *Applied Health Services Research,* Lexington, Mass.: Lexington Books, 1977.

73. Shonick, W. *Elements of Planning Area-Wide Personal Health Services.* St. Louis: C.V. Mosby, 1976, p. 1.

74. Shonick, W. 1976, p. 3.

75. Thompson, J.D. 1977, p. 168.

76. Thompson, J.D. 1977, p. 168.

77. Mac Stravic, R. *Determining Health Needs.* Ann Arbor, Mich.: Health Administration Press, 1978.

78. *Glossary of Health Planning Terms.* Midwest Center for Health Planning, G77/3, April 1977.

79. Warheit, G.J., Bell, R.A., and Schwab, J.J. *Planning for Change: Needs Assessment Approaches.* Rockville, Md.: National Institute of Mental Health, 1974.

80. Hargreaves, W.A., Attkisson, C.C., and Sorensen, J.E., *Resource Materials for Community Mental Health Program Evaluation.* Rockville, Md.: National Institute of Mental Health, 1977.

81. Warheit, G.J., Bell, R.A., and Schwab, J.J. 1974.
82. Muraco, W.A. The Lucas County mental health system: an evaluation and needs assessment planning report, Sept. 1977.
83. Feldstein, M.S. *The Rising Cost of Hospital Care.* Washington, D.C.: National Center for Health Services Research, Information Resources Press, 1971.
84. Mechanic, D. *The Growth of Bureaucratic Medicine: An Inquiry into the Dynamics of Patient Behavior and the Organization of Medical Care.* New York, 1976.
85. I wish to thank Eli Ginzberg, Osler Peterson, Jack Fein, Lee Friedman, William Niskanen, Marc Pauley, Otto Davis, and Merlin Du Val for their helpful comments on various drafts of this paper. Responsibility for the final version, however, is mine.

Conclusion

External environmental influences such as competition, economic disfunction, social values, client requirements, public mandates, and resource scarcity persistently play a role in shaping the inputs of an organization, its structure, and the courses of action available to managers within the organizational system. The purposes of this book have been twofold: one, to examine the nature of organization-environment relationships as identified by research, and two, to tie this emerging theory to the practicalities of management within a health care setting. Unfortunately, the theory does not readily translate into prescriptive courses of action for managers to follow. The open-systems approach, upon which the study of organization and environment is based, suggests that there are varying numbers of external factors influencing organization performance and that there are many routes to the same level of organization effectiveness.

Since the theoretical framework for understanding organization and environment is not developed to the point where direct application of research findings is feasible, the approach of this book has been to present a few key studies showing how the organization-environment relationship is being perceived, and then to offer examples of how people have implemented actions which enhance congruency between an organization and its environment.

The study of organization-environment relationships sensitizes managers to their responsibility of systematically and consistently perceiving and interpreting the present and future states of the external environment. As social interdependency continues to accelerate, managers will be increasingly called upon to analyze the different environment situations being encountered and to integrate the organization with its larger environment using sophisticated strategies. Beyond the immediate organization prob-

225

lems is the larger question of the ultimate purpose of the organization, which has yet to be fully addressed.

A statement by James Thompson indicates the challenge to be considered by society as we each work to make our own institutions serve the communities which support them:

Organizations and Societies

Complex organizations exist ultimately as agencies of their environments, acquiring resources in exchange for outputs and, in the final analysis, obtaining technologies from environments. But task environments never are as extensive as societies, and there can be pockets in which several organizations operate in mutual support in a network which is itself at odds with the larger society. The fact that organizations exist with the consent of their environments does not automatically subject them to societal control.

We have no assurance that modern societies have yet developed procedures for assessing the value of the kinds of organizations modern societies are developing, but because there are powerful instruments we can expect modern societies to exhibit considerable concern over the uses and abuses of complex organizations.

Modern societies must also struggle with questions of developing and allocating human capacities for the administration of complex organizations. Comparative analyses of the sources, capacities, and behavior of administrators in various societies are essential if we are to understand organizations in action, or to control them.[1]

REFERENCES AND NOTES

1. Thompson, J. *Organizations in Action.* New York: McGraw-Hill, 1967, p. 162.

CONTRIBUTING AUTHORS

Robert R. Alford

Chairman, Board of Studies in Sociology
University of California
Santa Cruz, California

Frank Baker, Ph. D.

Professor, Departments of Psychiatry,
Psychology, and Social and Preventive
Medicine
State University of New York at Buffalo
Buffalo, New York

Harry V. Berg, M.A.

Director, Staff Development
W. A. Foote Memorial Hospital, Inc.
Jackson, Michigan

Gwendolyn J. Buchanan, R.N., M.S.N.

Clinical Assistant Professor
University of Oklahoma College of Nursing
Oklahoma City, Oklahoma

Mary Bakszysz Bymel†

Daniel M. Harris, Ph.D.

Assistant Professor, Health Services Management
University of Missouri-Columbia
Columbia, Missouri

Stanley R. Ingman†

Fremont E. Kast, M.B.A., Ph.D.

Professor, Management and Organization
Graduate School of Business Administration
University of Washington
Seattle, Washington

John R. Kimberly

Professor
Yale School of Organization and Management
New Haven, Connecticut

Sol Levine†

Ellen Ramsey, R.N., B.S.

Clinical Instructor
Coordinator, Nurse Internship Program
Fresno Community Hospital
Fresno, California*

James E. Rosenzweig, M.B.A., Ph.D.

Professor, Management and Organization
Graduate School of Business Administration
University of Washington
Seattle, Washington

Stephen L. Schensul†

Ralph M. Stogdill†

James Thompson†

Sandra J. Weiss, R.N., D.N.Sc.

Assistant Clinical Professor
School of Nursing
University of California
San Francisco, California

Thomas M. Wernert

Executive Director
Lucas County Mental Health and Mental
Retardation Board
Toledo, Ohio

Paul E. White, Ph.D.

Professor and Chairman, Department of
Behavioral Sciences
School of Hygiene and Public Health
The Johns Hopkins University
Baltimore, Maryland

Aaron Wildavsky, Ph.D.

Professor, Political Science
University of California
Berkeley, California

---

*This information reflects the author's position when the selection included in this anthology was originally published
†No biographical information available

Index